CYTOGENETICS,
ENVIRONMENT
AND
MALFORMATION
SYNDROMES

The National Foundation – March of Dimes
Birth Defects: Original Article Series Volume XII, Number 5, 1976

CYTOGENETICS, ENVIRONMENT AND MALFORMATION SYNDROMES

The 1975 BIRTH DEFECTS CONFERENCE
Held at Kansas City, Missouri
Sponsored by the University of Kansas Medical Center
College of Health Sciences and Hospital
and The National Foundation – March of Dimes

Editors: **Daniel Bergsma, M.D.,** Vice President for Professional
Education, The National Foundation

R. Neil Schimke, M.D., Departments of Medicine and
Pediatrics, University of Kansas Medical Center,
Kansas City, Kansas

Assistant Editor: **Sue Conde Greene,** The National Foundation

ALAN R. LISS, INC., NEW YORK, N.Y.

To enhance medical communication in the birth defects field,
The National Foundation publishes the *Birth Defects Atlas and
Compendium*, an *Original Article Series*, *Syndrome Identification*,
a *Reprint Series* and provides a series of films and related brochures.

Further information can be obtained from:

Professional Education Department
The National Foundation — March of Dimes
1275 Mamaroneck Avenue
White Plains, New York 10605

Published by:

Alan R. Liss, Inc.
150 Fifth Avenue
New York, New York 10011

Received for publication February 12, 1976

Library of Congress Cataloging in Publication Data

Birth Defects Conference, Kansas City, Mo., 1975.
  Cytogenetics, environment, and malformation syndromes.

  (Birth defects original article series; v. 12, no. 5)
  Includes bibliographical references and indexes.
    1. Fetus — Abnormities and deformities — Congresses.
2. Teratogenic agents — Congresses. 3. Human chromosome
abnormalities — Congresses. 4. Deformities — Congresses.
I. Bergsma, Daniel. II. Schimke, R. Neil, 1935—
III. Kansas. University. College of Health Sciences
and Hospital. IV. National Foundation. V. Title.
VI. Series. [DNLM: 1. Abnormalities — Congresses.
2. Environment — Congresses. 3. Chromosome abnormalities
— Congresses. W1 BI966 v. 12 no. 5 / [QS675 B619 c
1975]]
RG626.B63 vol. 12, no. 5 [RG627] 616'.043'08s 76-20510
ISBN 0-8451-1004-7   [616'.043]

Printed in the United States of America

THE NATIONAL FOUNDATION is dedicated to the long-range goal of preventing birth defects. Our interim goal is to search for ways to ameliorate those birth defects which cannot be prevented.

As a part of our efforts to achieve these goals, we sponsor, or participate in, a variety of scientific meetings and symposia where all questions relating to birth defects are freely discussed. Through our professional education program we speed the dissemination of information by publishing the proceedings of these meetings and symposia. Now and then, in the course of these discussions, individual participants may express personal viewpoints which go beyond the purely scientific in nature and into controversial matters; abortion for example. It should be noted, therefore, that personal viewpoints about such matters will not be censored but obviously this does not constitute an endorsement of them by The National Foundation.

Table of Contents

Contributors   xi

Dedication to Virginia Apgar, M.D.
  L. Stanley James   xx

SECTION I: Birth Defects and the Environment

Birth Defect Surveillance in the Search for and Evaluation of
  Possible Human Teratogens
  Godfrey P. Oakley, Jr.   1

Latency of Defects
  Edmond A. Murphy   7

Radiation-Induced Malformations in Man
  Timothy Merz   19

Immediate and Long-Range Effects of Maternal Viral
  Infection in Pregnancy
  Janet B. Hardy   23

Embryopathy Associated With Oral Anticoagulant Therapy
  Judith G. Hall.   33

Cortical Cataracts Following Total Parenteral Nutrition
  J. Denis Catalano and James A. Monteleone   39

Sacral Agenesis and Associated Anomalies
  *Jeanette N. Israel, Donald W. Day, Asna Hirschmann and
  George F. Smith*   45

Monozygotic Twinning and the Duhamel Anomalad
  (Imperforate Anus to Sirenomelia): A Nonrandom Association
  Between Two Aberrations in Morphogenesis
  David W. Smith, Catherine Bartlett and Lyle M. Harrah   53

Aberrant Tissue Bands and Craniofacial Defects
  *Diane L. Broome, Allan J. Ebbin, August L. Jung,
  L. Richard Feinauer and Michael Madsen*   65

Selected Abstracts   81

SECTION II: Cytogenetic Disorders

Usefulness of Chromosome Catalog in Delineating New
  Syndromes
  Digamber S. Borgaonkar, Yves E. Lacassie and Claude Stoll   87

E Trisomy Phenotype Associated With Small Metacentric
Chromosome and a Familial Y-22 Translocation
Kenneth W. Dumars, Gayle Fialko and Eunice Larson . . . . . .　97

Ring Y Chromosome Without Mosaicism
Miriam G. Wilson, Ronald B. Stein and Joseph W. Towner . .　105

Tertiary Trisomy 14: Is There a Syndrome?
*Sergio D. J. Pena, M. Ray, P. J. McAlpine, C. Ducasse,
J. Briggs and John L. Hamerton* . . . .　113

Partial Trisomy of Chromosome 14: (+14q−)
Gentry W. Yeatman and Vincent M. Riccardi . . . .　119

Deletion of 11q: Report of Two Cases and a Review
*Steven A. Larson, Gentry W. Yeatman
and Vincent M. Riccardi* . . . .　125

Terminal (1)(q43) Long-Arm Deletion of Chromosome No. 1
in a Three-Year-Old Female
Carl B. Mankinen, Joseph W. Sears and Victor R. Alvarez . . .　131

X Short-Arm Deletion Gonadal Dysgenesis in Two Sibs
*John R. Davis, M. Wayne Heine, Raymond F. Graap,
Elmer S. Lightner and Harlan R. Giles* . . . .　137

Trisomy C and Cystic Dysplasia of Kidneys, Liver and Pancreas
John D. Blair . . . .　139

9pter→p22 Deletion Syndrome: A Case Report
*Penelope W. Allderdice, Walter D. Heneghan and
Emerita T. Felismino* . . . .　151

The 9p− Syndrome
Omar S. Alfi, George N. Donnell and Anna Derencsenyi . . . .　157

Partial Monosomy and Partial Trisomy for Different Segments
of Chromosome 13 in Several Individuals of the Same Family
*Robert S. Wilroy, Jr., Robert L. Summitt, Paula Martens,
W. Manford Gooch, III, Carol Hood and Winfred Wiser.* . . . . .　161

Ambiguous Genitalia and Mental Retardation Associated
With a Translocation 46,XY,t(9;10) and a Deletion in 9q
*Edmund C. Jenkins, R. S. K. Chaganti, Lorraine Wilbur
and James German, III* . . . .　169

Selected Abstracts . . . .　175

SECTION III: Malformation Syndromes

Peters Anomaly With Pulmonary Hypoplasia
Marilyn J. Bull and Jules L. Baum 181

Peripheral Pulmonary Cystic Disease in Sibs
Elizabeth J. Ives, Maria Darja and Susana Geist 187

A Syndrome of Ankylosis, Facial Anomalies and
Pulmonary Hypoplasia Secondary to Fetal
Neuromuscular Dysfunction
*Alan D. Mease, Gentry W. Yeatman, Gary Pettett and
Gerald B. Merenstein* 193

Syndrome of Camptodactyly, Multiple Ankyloses, Facial
Anomalies and Pulmonary Hypoplasia — Further Delineation
and Evidence for Autosomal Recessive Inheritance
Sergio D. J. Pena and Mohamed H. K. Shokeir 201

Accelerated Skeletal Maturation Syndrome With Pulmonary
Hypertension
*Jane C. S. Perrin, Edgardo Arcinue, William H. Hoffman,
Harold Chen and Joseph O. Reed* 209

A Syndrome Manifested by Brittle Hair With Morphologic and
Biochemical Abnormalities, Developmental Delay and
Normal Stature
*Amir I. Arbisser, Charles I. Scott, Jr., R. Rodney Howell,
Poen S. Ong and Hollace L. Cox, Jr.* 219

Familial Aortic Dissection With Iris Anomalies — A New
Connective Tissue Disease Syndrome?
David Bixler and Ray M. Antley. 229

A Possible New Mental Retardation Syndrome
Jessica G. Davis and Charlotte Lafer 235

Focal Palmoplantar and Marginal Gingival
Hyperkeratosis — A Syndrome
Robert J. Gorlin 239

Monosuperocentroincisivodontic Dwarfism
*Elizabeth B. Rappaport, Robert Ulstrom
and Robert J. Gorlin* 243

Studies of Malformation Syndromes in Man XXXXII: A
Pleiotropic Dominant Mutation Affecting Skeletal, Sexual
and Apocrine-Mammary Development
Philip D. Pallister, Jürgen Herrmann and John M. Opitz 247

Hallux Syndactyly — Ulnar Polydactyly — Abnormal Ear
   Lobes: A New Syndrome
   Michael J. Goldberg and Hermine M. Pashayan 255

X-Linked Syndrome of Congenital Ichthyosis, Hypogonadism,
   Mental Retardation and Anosmia
   Jane C. S. Perrin, Judi Y. Idemoto, Juan F. Sotos,
   William F. Maurer and Arthur G. Steinberg 267

Congenital Microcephaly, Hiatus Hernia and Nephrotic
   Syndrome: An Autosomal Recessive Syndrome
   Lawrence R. Shapiro, Peter A. Duncan,
   Peter B. Farnsworth and Martin Lefkowitz 275

Polydactyly With Triphalangeal Thumbs, Brachydactyly,
   Camptodactyly, Congenital Dislocation of the Patellas,
   Short Stature and Borderline Intelligence
   Burhan Say, Eugene Feild, James G. Coldwell,
   Larry Warnberg and Metin Atasu 279

Recurrent Dislocation of the Patella Versus Generalized
   Joint Laxity
   Steven D. Shapiro, Ronald J. Jorgenson and Carlos F. Salinas 287

An X-Linked Form of Cutis Laxa Due to Deficiency of
   Lysyl Oxidase
   Peter H. Byers, A. Sampath Narayanan, Paul Bornstein
   and Judith G. Hall 293

Generalized Osseous Abnormalities in the Marshall Syndrome
   James J. O'Donnell, Shari Sirkin and Bryan D. Hall 299

Deafness and Vitiligo
   Theodore F. Thurmon, Jennifer Jackson
   and Cynthia G. Fowler 315

Tumoral Calcinosis and Engelmann Disease
   Theodore F. Thurmon and Jennifer Jackson 321

Bilateral Hammertoes: A New Dominant Disorder
   William B. Reed and Les Schlesinger 327

Selected Abstracts 329

Author Index 335

Subject Index 337

Contributors

Omar S. Alfi, M.D.
Division of Medical Genetics
Children's Hospital of Los Angeles
Los Angeles, CA 90054

Penelope W. Allderdice, Ph.D.
Memorial University of Newfoundland
St. John's, Newfoundland, Canada
A1C 5S7

Victor R. Alvarez, M.D.
Westgate Hospital
4401 I Highway 35 N
Denton, TX 76201

Ray M. Antley, M.D.
Director of Medical Genetics
Methodist Hospital
Indianapolis, IN
46202

Amir I. Arbisser, M.D.
Fellow in Genetics
Department of Pediatrics
University of Texas Medical School
Houston, TX 77025

Edgardo Arcinue, M.D.
Children's Hospital of Michigan
3901 Beaubien
Detroit, MI 48201

Metin Atasu, Ph.D.
Division of Clinical Genetics
Hacettepe University
Ankara, Turkey

Catherine Bartlett
11 Woodlawn Avenue
Albany, NY 12208

Jules L. Baum, M.D.
Professor of Ophthalmology
Tufts-New England Medical Center
Boston, MA 02111

David Bixler, D. D. S., Ph.D.
Chairman and Professor
Department of Oral-Facial Genetics
Indiana University School of
Dentistry
Indianapolis, IN 46202

John D. Blair, M.D.
Director
Department of Pathology
Cardinal Glennon Memorial
Hospital for Children
St. Louis, MO 63104

Digamber S. Borgaonkar, Ph.D.
Associate Professor of Medicine
Division of Medical Genetics
The Johns Hopkins
Medical Institutions
Baltimore, MD 21205

Paul Bornstein, M.D.
Professor of Biochemistry and
Medicine
University of Washington School
of Medicine
Seattle, WA 98195

J. Briggs, M.D., F.R.S.P. (C)
Winnipeg Clinic
St. Mary & Vaughan
Winnipeg, Manitoba, Canada

Diane L. Broome, M.D.
Department of Pediatrics
Genetics Division
Los Angeles County—USC Medical
Center
Los Angeles, CA 90033

Marilyn J. Bull, M.D.
Assistant Professor of Pediatrics
Center for Birth Defect Evaluation
and Genetic Counseling
Boston Floating Hospital
Boston, MA 02111

Peter H. Byers, M.D.
Senior Fellow, Biochemistry
and Medicine
University of Washington School
of Medicine
Seattle, WA 98195

J. Denis Catalano, M.D.
Assistant Clinical Professor of
Ophthalmology and Pediatrics
St. Louis University School
of Medicine
St. Louis, MO 63104

R. S. K. Chaganti, Ph.D.
Associate Investigator
Laboratory of Human Genetics
The New York Blood Center
New York, NY 10021

Harold Chen, M.D.
Children's Hospital of Michigan
Detroit, MI 48201

James G. Coldwell, M.D.
Associate Director
Children's Medical Center
Tulsa, OK 74135

Hollace L. Cox, Jr., Ph.D.
Postdoctoral Fellow
Laboratory for Bioanalytical
Physics
University of Texas System
Cancer Center
Houston, TX 77025

Maria Darja, M.D.
Associate Professor of Pathology
University of Saskatchewan
Saskatoon, Saskatchewan, Canada

Jessica G. Davis, M.D.
Director
Child Development and
Genetics Program
North Shore University Hospital
Manhasset, NY 11030

John R. Davis, M.D.
Professor
Department of Pathology
Arizona Medical Center
Tucson, AZ 85724

Donald W. Day, M.D.
Assistant Professor of Pediatrics
Rush-Presbyterian-St. Luke's
Hospital
Chicago, IL 60612

Anna Derencsenyi
Chief Technologist
Cytogenetics Laboratory
Children's Hospital of Los Angeles
Los Angeles, CA 90027

George N. Donnell, M.D.
Professor of Pediatrics
Division of Medical Genetics
Children's Hospital of Los Angeles
Los Angeles, CA 90027

C. Ducasse, M.D., F.B.C.P. (C)
Department of Genetics
Health Sciences Centre,
Children's Centre
Winnipeg, Manitoba, Canada
R3E OW1

Kenneth W. Dumars, M.D.
Associate Professor and Chief
Division of Developmental
Disabilities and Clinical Genetics
University of California College
of Medicine
Irvine, CA 92664

Peter A. Duncan, M.D.
Coordinator
Division of Pediatrics
Birth Defects Center
Westchester County Medical Center
Valhalla, NY 10595

Allan J. Ebbin, M.D.
Department of Pediatrics
Genetics Division
Los Angeles County–USC
    Medical Center
Los Angeles, CA 90033

Peter B. Farnsworth, M.D.
Director
Division of Pediatrics
Westchester County Medical Center
Valhalla, NY 10595

Eugene Feild, M.D.
Orthopedic Surgeon
St. Francis Hospital
Tulsa, OK 74136

L. Richard Feinauer, M.D.
Department of Pediatrics
University of Utah
Salt Lake City, UT 84112

Emerita T. Felismino, M.D.
Central Newfoundland Hospital
Grand Falls, Newfoundland, Canada

Gayle Fialko
Research Associate
Department of Pediatrics
University of California College
    of Medicine
Irvine, CA 92264

Cynthia G. Fowler, M.D.
Researcher
Department of Pediatrics
Louisiana Heritable Disease Center
Louisiana State University School
    of Medicine
New Orleans, LA 70112

Susana Geist, M.D.
Staff Pathologist
Pasqua Hospital
Regina, Saskatchewan, Canada

James German, III, M.D.
Senior Investigator and Director
Laboratory of Human Genetics
The New York Blood Center
New York, NY 10021

Harlan R. Giles, M.D.
Director
Genetic Information and
    Perinatal Reference Unit
Arizona Medical Center
Tucson, AZ 85724

Michael J. Goldberg, M.D.
Assistant Professor of
    Orthopedic Surgery
Tufts–New England Medical Center
Boston, MA 02111

W. Manford Gooch, III, M.D.
Associate Professor of Pathology
    and Pediatrics
University of Tennessee Center
    for the Health Sciences
Memphis, TN 38163

Robert J. Gorlin, D.D.S.
Professor and Chairman
Division of Oral Pathology
University of Minnesota School
    of Dentistry
Minneapolis, MN 55455

Raymond F. Graap, M.D.
Pima County Hospital
Tucson, AZ 85709

Bryan D. Hall, M.D.
Associate Director of Birth
    Defects Center
Department of Pediatrics
University of California
San Francisco, CA 94143

Judith G. Hall, M.D.
Director
Medical Genetics
The Children's Orthopedic Hospital
and Medical Center
Seattle, WA 98105

John L. Hamerton, D.Sc.
Director
Division of Genetics
Health Sciences Centre,
Children's Centre
Winipeg, Manitoba, Canada
R3E OW1

Janet B. Hardy, M.D., C.M.
Professor
Department of Pediatrics
The Johns Hopkins Medical
Institutions
Baltimore, MD 21205

Lyle M. Harrah
Research Librarian
University of Washington
Health Sciences Library
Seattle, WA 98195

M. Wayne Heine, M.D.
Department of Obstetrics and
Gynecology
University of Arizona College
of Medicine
Tucson, AZ 85724

**Walter D. Heneghan, M.B., B.Ch.
F.R.C.P. (C)**
Janeway Child Health Center
St. John's, Newfoundland, Canada

Jürgen Herrmann, M.D.
Assistant Professor of Pediatrics
University of Wisconsin Center
for Health Sciences and
Medical School
Madison, WI 53706

Asna Hirschmann, M.D.
Clinical Associate Professor
of Pediatrics
St. Joseph's Hospital
Chicago, IL 60657

William H. Hoffman, M.D.
Children's Hospital of Michigan
Detroit, MI 48201

Carol Hood, B.A.
Senior Research Technician
University of Tennessee Center
for the Health Sciences
Memphis, TN 38163

R. Rodney Howell, M.D.
Chairman
Department of Pediatrics
University of Texas Medical School
Houston, TX 77025

Judi Y. Idemoto, M.S.W.
Department of Pediatrics
Genetic Social Worker
Cleveland Metropolitan Hospital
Cleveland, OH 44109

Jeanette N. Israel, M.D.
Assistant Professor of Pediatrics
Rush-Presbyterian-St. Luke's
Hospital
Chicago, IL 60612

Elizabeth J. Ives, F.R.C.P. (C)
Professor and Head
Department of Pediatrics
University Hospital
Saskatoon, Saskatchewan, Canada

Jennifer Jackson, B.S.
Genetics Associate
Department of Pediatrics
Louisiana Heritable Disease Center
Louisiana State University School
of Medicine
New Orleans, LA 70112

L. Stanley James, M.D.
Professor of Pediatrics
College of Physicians and
  Surgeons of Columbia
  University
New York, NY 10032

Edmund C. Jenkins, Ph.D.
Head
Department of Cytogenetics
The New York State Institute for
  Basic Research in Mental
  Retardation
Staten Island, NY 10314

Ronald J. Jorgenson, D.D.S.
Assistant Professor
Department of Pediatric Dentistry
College of Dental Medicine
Medical University of
  North Carolina
Charleston, SC 29401

August L. Jung, M.D.
Department of Pediatrics
University of Utah
Salt Lake City, UT 84112

Yves E. Lacassie, M.D.
Postdoctoral Fellow
Department of Medicine
The Johns Hopkins
  Medical Institutions
Baltimore, MD 21205

Present Address
Department of Pediatrics
Louisiana State University
  Medical Center
New Orleans, LA 70112

Charlotte Lafer, M.D.
Assistant Clinical Professor of
  Pediatrics
Cornell University College of
  Medicine
Manhasset, NY 11030

Eunice Larson
Pathologist
Department of Pathology
Long Beach Memorial Children's
  Hospital
Long Beach, CA 90801

Steven A. Larson, Cpt., M.C., U.S.A.
Department of Pediatrics
Fitzsimons Army Medical Center
Denver, CO 80240

Martin Lefkowitz, M.D.
Department of Pathology
Westchester County Medical Center
Valhalla, NY 10595

Elmer S. Lightner, M.D.
Department of Pediatrics
University of Arizona College
  of Medicine
Tucson, AZ 85724

Michael Madsen
Department of Pediatrics
University of Utah
Salt Lake City, UT 84112

Carl B. Mankinen, Ph.D.
Director of Cytogenetics
Genetics of Screening and
  Counseling Service
Denton State School
Denton, TX 76202

Paula Martens, M.S.
Senior Research Assistant
University of Tennessee Center
  for the Health Sciences
Memphis, TN 38163

William F. Maurer, M.D.
Fellow in Pediatric Endocrinology
Department of Pediatrics
Ohio State University
Columbus, OH 43210

P. J. McAlpine, Ph.D.
Assistant Professor of Paediatrics
Department of Genetics
Health Sciences Centre,
Children's Centre
Winnipeg, Manitoba, Canada
R3E OW1

Alan D. Mease, Cpt., M.C., U.S.A.
Fellow
Pediatric Hematology/
Oncology Service
Walter Reed Army Medical Center
Washington, DC 20021

**Gerald B. Merenstein, Lt.C.,
M.C., U.S.A.**
Assistant Chief
Department of Pediatrics
Fitzsimons Army Medical Center
Denver, CO 80240

Timothy Merz, Ph.D.
Chairman and Professor
Department of Radiation Biology
Virginia Commonwealth University
Medical College of Virginia
Richmond, VA 23298

James A. Monteleone, M.D.
Associate Professor of Pediatrics
St. Louis University School
of Medicine
St. Louis, MO 63103

Edmond A. Murphy, M.D.
Director
Division of Medical Genetics
The Johns Hopkins Medical
Institutions
Baltimore, MD 21205

A. Sampath Narayanan, Ph.D.
Research Assistant Professor
Department of Pathology
University of Washington School
of Medicine
Seattle, WA 98195

Godfrey P. Oakley, Jr., M.D.
Chief
Etiologic Studies Section
Birth Defects Branch
Bureau of Epidemiology
Center for Disease Control
Atlanta, GA 30333

James J. O'Donnell, M.D.
Fellow-in-Genetics
Departments of Pediatrics and
Ophthalmology
University of California
San Francisco, CA 94143

Poen S. Ong, Ph.D.
Associate Professor of Physics
Laboratory for Bioanalytical
Physics
University of Texas System
Cancer Center
Houston, TX 77025

John M. Opitz, M.D.
Director
Wisconsin Clinical Genetics Center
University of Wisconsin Center for
Health Sciences and Medical
School
Madison, WI 53706

Philip D. Pallister, M.D.
Clinical Director
Boulder River School and Hospital
Boulder, MT 59632

Hermine M. Pashayan, M.D.
Associate Professor of Pediatrics
Tufts-New England Medical Center
Boston, MA 02111

Sergio D. J. Pena, M.D.
Postdoctoral Fellow
Department of Genetics
Health Sciences Centre,
Children's Centre
Winnipeg, Manitoba, Canada
R3E OW1

Jane C. S. Perrin, M.D.
Children's Hospital of Michigan
Detroit, MI 48201

Gary Pettett, Maj., M.C., U.S.A.
Neonatology Fellow
Department of Pediatrics
Fitzsimons Army Medical Center
Denver, CO 80240

Elizabeth B. Rappaport, M.D.
Resident
Montreal Children's Hospital
Montreal, Quebec, Canada H3H 1P3

M. Ray, Ph.D.
Assistant Professor of Paediatrics
Department of Genetics
Health Sciences Centre,
    Children's Centre
Winnipeg, Manitoba, Canada
    R3E OW1

Joseph O. Reed, M.D.
Children's Hospital of Michigan
Detroit, MI 48201

William B. Reed, M.D.
Clinical Professor of Dermatology
University of California
Irvine, CA 92664

Vincent M. Riccardi, M.D.
Director
Genetics Unit
University of Colorado
    Medical Center
Denver, CO 80220
Present Address
    Genetics Unit
    Milwaukee Children's Hospital
    Milwaukee, WI 53233

Carlos F. Salinas, D.D.S.
Visiting Instructor
Department of Pediatric Dentistry
College of Dental Medicine
Medical University of
    South Carolina
Charleston, SC 29401

Burhan Say, M.D.
Director of the Laboratories
Children's Medical Center
Tulsa, OK 74135

Les Schlesinger, M.D.
Radiologist
St. Joseph's Medical Center
Burbank, CA 91506

Charles I. Scott, Jr., M.D.
Director
Medical Genetics Clinic
University of Texas Medical School
Houston, TX 77025

Joseph W. Sears, M.D.
Medical Coordinator
Genetics Screening and
    Counseling Service
Denton State School
Denton, TX 76202

Lawrence R. Shapiro, M.D.
Associate Clinical Professor
Pediatrics and Pathology
New York Medical College
Valhalla, NY 10595

Steven D. Shapiro, M.S.
Research Associate
Department of Pediatric Dentistry
College of Dental Medicine
Medical University of
    South Carolina
Charleston, SC 29401

Mohamed H. K. Shokeir, M.D., Ph.D.
Department of Genetics
Health Sciences Centre,
    Children's Centre
Winnipeg, Manitoba, Canada
    R3E OW1

Shari Sirkin, M.D.
Pediatric Intern
Vanderbilt University
Nashville, TN 37235

David W. Smith, M.D.
Professor in Pediatrics
RR234 Health Sciences RD-20
University of Washington School
of Medicine
Seattle, WA 98195

George F. Smith, M.D.
Professor of Pediatrics
Rush-Presbyterian-St. Luke's
Hospital
Chicago, IL 60612

Juan F. Sotos, M.D.
Director of Endocrinology
Department of Pediatrics
Ohio State University
Columbus, OH 43210

Ronald B. Stein, M.D.
Associate Clinical Professor of
Medicine
Los Angeles County–USC
Medical Center
Los Angeles, CA 90033

Arthur G. Steinberg, Ph.D.
Francis Hobart Herrick Professor
of Biology
Case Western Reserve University
Cleveland, OH 44106

Claude Stoll, M.D.
Postdoctoral Fellow
Department of Medicine
The Johns Hopkins
Medical Institutions
Baltimore, MD 21205
Present Address
Service de Pediatrie IV
Center Hospitalier
Universitaire
Strasbourg, France

Robert L. Summitt, M.D.
Professor
Departments of Pediatrics, Anatomy
and Child Development
University of Tennessee Center for
the Health Sciences
Memphis, TN 38163

Theodore F. Thurmon, M.D.
Director
Department of Pediatrics
Louisiana Heritable Disease Center
Louisiana State University School
of Medicine
New Orleans, LA 70112

Joseph W. Towner, Ph.D.
Assistant Professor, Genetics
Division
Department of Pediatrics
Los Angeles County–USC
Medical Center
Los Angeles, CA 90033

Robert Ulstrom, M.D.
Professor of Pediatrics
University of Minnesota School
of Medicine
Minneapolis, MN 55455

Larry Warnberg, M.A.
Psychologist
Children's Medical Center
Tulsa, OK 74135

Lorraine Wilbur, B.S.
Assistant Research Scientist
Department of Cytogenetics
The New York State Institute for
Basic Research in Mental
Retardation
Staten Island, NY 10314

Robert S. Wilroy, Jr., M.D.
Associate Professor of Pediatrics
and Child Development
University of Tennessee Center for
the Health Sciences
Memphis, TN 38163

Miriam G. Wilson, M.D.
Chief
Department of Pediatrics
Genetics Division
Los Angeles County–USC
Medical Center
Los Angeles, CA 90033

Winfred Wiser, M.D.
Professor of Obstetrics &
Gynecology
University of Tennessee Center for
the Health Sciences
Memphis, TN 38163

Gentry W. Yeatman, Maj.,
M.C., U.S.A.
Resident in Pediatrics
Fitzsimons Army Medical Center
Denver, CO 80240
Present Address
Department of Pediatrics
Martin Army Hospital
Fort Benning, GA 31905

Dedication to Virginia Apgar, MD

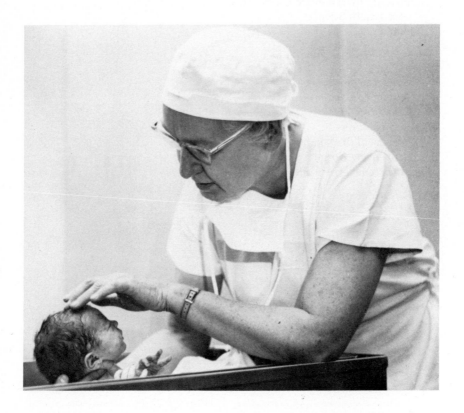

So great was her influence that eighteen months after her death, Virginia Apgar's memory is still vivid and her presence, real.

She had many special talents. Her approach to problems was direct and forthright. Because she was never embarrassed herself, she could deal with sensitive or personal problems in a realistic and practical manner. This enabled her to bring congenital anomalies into the open — the first step in planning a research or preventive program. She carried a small bottle containing an 8-week-old fetus, as a teaching device, in her purse. The fetus had a failure of closure of the neural tube and therefore had a visible defect which could clearly be seen at that early age. The fetus was even named Billy, and at a dinner meeting of the Spina Bifida

Birth Defects: Original Article Series, Volume XII, Number 5, pages xx—xxi
© 1976 The National Foundation

Association of America, where Virginia was the principal speaker, she produced Billy. He was handed around and examined by all the parents. No one was shocked. All were fascinated and intrigued. It was a perfect educational vehicle.

She achieved her greatest visibility in later years in her drive to educate the whole country about the need for early detection of birth defects. She almost never turned down an invitation to speak, no matter how small or apparently insignificant the group, and her life became one long juggling act to fit speeches and site visits, professional consultations and chapter meetings, media interviews and international congresses into her impossible schedule. She was the finest ambassador The National Foundation ever had. Undoubtedly, she lifted birth defects from the closet and put them firmly on the map.

As an investigator, Virginia was completely dedicated to seeking out new information that would improve the quality of life. Failure to do this was, to her, almost immoral. When Virginia was there, there was a sense of excitement; and in some way, she made what you were doing seem important. She encouraged constantly, and if she believed in your ability, you just had to do it. Old and new ideas were constantly reexamined with objectivity. She had an almost clairvoyant ability to pick out the essentials of a problem and an even greater ability to discard the claptrap. Her own achievements she considered very minor and unimportant. But she did see herself as a catalyst and took the greatest pride in bringing people with complementary interests and talents together to achieve a specific goal.

Virginia was completely devoid of vanity and conceit. Her modesty and humility were just natural qualities that pervaded every aspect of her being and her life. Her inspiration lives on as a guide to us all.

L. Stanley James, MD
Professor of Pediatrics
College of Physicians & Surgeons
of Columbia University

SECTION I:
BIRTH DEFECTS AND THE ENVIRONMENT

Birth Defect Surveillance in the Search for and Evaluation of Possible Human Teratogens

Godfrey P. Oakley, Jr., MD

This section is devoted to birth defects and the environment. You will read status reports on many teratogens or suspected teratogens. By the end of the section, it should be even clearer to you than perhaps it now is that the task of finding environmental agents responsible for birth defects is slow and difficult and usually involves a variety of people and approaches. It is seldom, if ever, that a single observation immediately results in the establishment of an environmental agent as a teratogen. My topic concerns the role that birth defect surveillance and epidemiologic approaches may play in the process of establishing and evaluating teratogens or suspected teratogens. In order to show this role in proper perspective, I have decided first to present a brief overview of the problem of finding human teratogens.

There are 3 categories in which we perceive human teratogens. Some agents are generally accepted as being known teratogens by all of us, eg rubella and thalidomide. There are other teratogens that, at this point, are only suspected; three years ago, warfarin would have been in this category. There is also a group of teratogens of unknown size that are not seriously expected to be teratogens. Some in this group are now causing birth defects and others, because of no current exposure, are not. Among those causing birth defects, some began to affect human development only in recent times, while others may have begun with Adam and Eve. Our primary goals are to keep human embryos from being exposed to teratogens not now causing disease and to place all teratogens now causing disease into the known category. An ancillary goal is to minimize the number of environmental agents incorrectly implicated as a teratogen.

How do we go about preventing new agents from causing birth defects? Over the years since thalidomide, a great deal of time has been spent discussing this

Birth Defects: Original Article Series, Volume XII, Number 5, pages 1–6

problem with regard to a single group of agents — drugs. The main action taken has been to require premarketing teratologic screening studies in animals. Since this began, there has been no major epidemic of birth defects observed in this country. I would be cautious, however, about accepting a cause and effect relationship; there were no drug-related birth defect epidemics in this country before the testing. Furthermore, the discovery of drug teratogens that do not cause obvious epidemics is a slow process. Witness the 40 years between marketing and the first suspicions that diphenylhydantoin is teratogenic. It is possible that a number of drugs, marketed after the animal screens were begun, have teratologic properties in humans that are hard to discover.

Another approach to this problem would be to follow and study systematically the outcome of pregnancy in the first group of pregnant women exposed to the drugs. Such studies are difficult to do and are expensive. As an example of the problems, one would need to study approximately 20,000 women exposed in the first trimester to have a reasonably good chance of finding in the offspring a twofold increase in the incidence of common serious malformations. For rarer defects the numbers to study would have to be even larger. Such detailed study of large groups is not done for drugs now, nor is it likely to be done on a large scale in the future. The point that I am trying to make is that present day technology does not guarantee that drugs or other chemicals that may come into the environment are free of teratogenic potential. We will, therefore, have the task of trying to find these new teratogens in a sea of old recognized ones.

How do we go about finding active teratogens? Past experiences have shown us that recognition of an epidemic of birth defects led to investigations that eventually established a teratogen as the cause of the epidemic. Likewise, teratogens have been found when etiologic questions were asked about birth defects not being seen in epidemic proportions. Let us look at teratogen discovery in these 2 settings.

Figure 1 represents steps usually involved in solving the mystery of an epidemic. The left extreme represents the time the first case of a birth defect is caused by the exposure; the far right is the point when the epidemic is curtailed. Conceptually, there are 2 main time intervals that I would like to call to your attention. The first is between A and B, which is the time between the first case and the recognition of the epidemic. The second is the time between B and G, representing the time between the recognition of the epidemic and the institution of preventive measures. The time span covered by these intervals is usually measured in years and is quite variable. Furthermore, if one cuts the interval between the time the agent is introduced and the epidemic is discovered, one does not necessarily change the interval between the discovery and ending of the epidemic. What goes on after an epidemic is discovered? Basically, a search for the agent begins in a phase I have labeled the Rorschach effect; it is a time of hypothesis generation. I have labeled it Rorschach because the kinds of

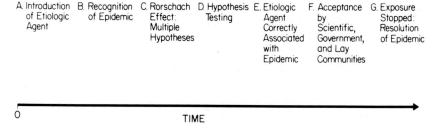

A. Introduction of Etiologic Agent   B. Recognition of Epidemic   C. Rorschach Effect: Multiple Hypotheses   D. Hypothesis Testing   E. Etiologic Agent Correctly Associated with Epidemic   F. Acceptance by Scientific, Government, and Lay Communities   G. Exposure Stopped: Resolution of Epidemic

Fig. 1. Events between introduction of new agent causing epidemic disease and ending of epidemic.

hypotheses one generates usually reflect the area in which one works. The infectious disease people propose an infectious etiology, nutritionists will see it as a result of a deficiency and so forth.

Some people will think enough of their hypotheses to test them. Gradually, the correct hypothesis will be tested and the others begin to fall away. Eventually, the evidence persuades the government, scientific and lay communities to take preventive measures. The interval between recognition of the epidemic and solution varies — with thalidomide it was about 2 years; for retrolental fibroplasia it was 15 years. Although epidemics in general have led to the discovery of the etiology of many diseases, epidemics of birth defects have been rare. Hence, only a few teratogens have been discovered in an epidemic setting.

Human teratogens have usually been found from observations in nonepidemic times. Usually a clinical observation of a clinical peculiarity of the case has led to hypothesis generation based on a careful history. Again, the interval between the time the agent began to cause a birth defect and the time a clinical observation leads to a correct etiologic hypothesis varies and may span centuries — consider the fetal alcohol syndrome. The interval between the clinically raised hypothesis and acceptance is also varied — consider the controversy that still surrounds the diphenylhydantoin situation.

Where can birth defect surveillance be of use in this process of discovering teratogens? The widespread use of a serious new teratogen would be expected to cause a change in the incidence of the birth defect it was engendering. The main raison d'être for birth defect surveillance systems is to monitor the incidence of birth defects in an attempt to detect changes in incidence. In other words, monitoring is directed at reducing the interval between the time an agent begins to cause the defects and the time the problem is noted. It does not necessarily affect the time between recognition of a problem and its solution.

There are now several examples from data sets analyzed in Atlanta and from those analyzed abroad where monitoring systems have picked up epidemics that were later explained. Fortunately, these were what we call administrative epidemics. The sudden aggressive reporting of a poorly described condition and changes in coding personnel and/or coding criteria have accounted for such administrative epidemics. We feel if there were major shifts in the incidence of a monitored birth defect, we would be able to see it. But, like the premarketing animal studies, the monitoring programs have, fortunately, not been tested by a major epidemic.

What are some of the problems with the monitoring? First of all, if one looks at a large number of birth defect categories and measures their incidence frequently, one will have statistically significant changes in incidence due to random variation. One spends a good deal of time trying to sort through these to find a real problem. So, monitoring systems can have trouble with false-positive results, and likewise, with false-negatives. For example, in the incidence data from Atlanta, there is no clue that warfarin, Dilantin or alcohol are teratogenic. In addition, the use of these agents antedated the monitoring programs. False-negative reports may arise because a teratogen either affects only a small percentage of exposed embryos and/or because exposure is quite limited. For example, we have recently found that if 5% of the pregnant women were exposed to an agent that increased the incidence of cleft lip with or without cleft palate fourfold, the incidence would be expected to change only 17% — a level that would be difficult to discern from background rates. If the risk for the defect was increased 20-fold, then the monitoring would probably detect a change.

Monitoring data can, however, be of use in the evaluation of suspect agents. In fact, its main use has been in this area. Of concern to all of us is whether the hormones in contraceptive pills are teratogenic. The most recent concern is whether they may cause limb-reduction deformities (1). Significant use of the pill also antedated any of the data sets that we have in Atlanta. In Birmingham, England, however, the incidence of birth defects has been monitored since 1950 (2). It was recently reported from there that there had been no secular trend in the incidence of reduction deformities. No single datum can establish or rule out a hypothesis. These monitoring data do show, however, that no major epidemic exists from this exposure.

Monitoring data, while not being able per se to reject an etiologic hypothesis, may rather rapidly be able to give clinically useful information. You may recall that about 2 years ago spray adhesives were taken off the market because of data linking their use to chromosome breakage and birth of children with multiple defects. Exposed women sought counseling. When monitoring data were linked to a quick investigation to determine exposure patterns, it was possible to put an upper limit on the magnitude of the risk, if there was any risk at all.

Neither the total incidence of birth defects nor the incidence of infants with multiple birth defects increased in Atlanta (Fig. 2). There were also no significant

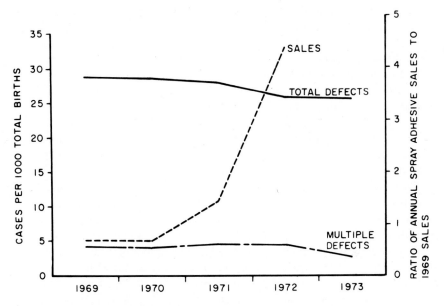

Fig. 2. Trends in birth defect incidence and spray adhesive sales in metropolitan Atlanta, 1969–1973. (From Heath, C. W., Jr., Flynt, J. W., Jr., Oakley, G. P., Jr. and Falek, A.: The rôle of birth defect surveillance in control of fetal environmental hazards. Postgrad. Med. J. 51:69–77, 1975, with permission.)

secular trends for any birth defect rubric monitored in Atlanta. Concurrently, sales for spray adhesives had increased dramatically in the area. We also found out that about 5% of pregnant women who had delivered in Atlanta had been exposed to spray adhesives. Knowing that there were 30,000 births a year in Atlanta and knowing that virtually all the birth defect rubrics occurred with an incidence of 1 in 1000 or less, we were able to predict that the maximum risk for a birth defect to an exposed embryo was about 5%. While such data did not disprove the hypothesis, it did give a clinically useful figure to use for discussion during abortion counseling (3).

The data collected for monitoring also can be used in other ways. If the cases are entered in a registry, they can be used to look for or to evaluate suspected teratogens. In Atlanta, the cases are entered in a registry. A subset of the parents of infants with defects listed on the registry, primarily those with serious defects that are consistently reported, are routinely interviewed. The purpose of the interview is to collect information about the prenatal history. Some of the questions are designed to test specific hypotheses, like the role of anticonvulsants in the etiology of these defects. Other questions are asked in the hope of generating hypotheses about either old or new teratogens. We are currently in the midst of analyzing data on maternal drug use from the first 700 of these interviews. Our data are consistent with the observation that women with seizure disorders who take diphenylhydantoin are at an increased risk of having children with cleft lip and cleft palate. Other analyses are beginning to raise hypotheses that need testing.

Parents of registry cases are an everready source that can be used to investigate a serious hypothesis raised by others. The Atlanta data have been used to evaluate such hypotheses. In 1971 when the hypothesis was raised that the tricyclic antidepressants caused limb-reduction deformities, we were not routinely interviewing mothers of children with this defect. Mothers of infants listed in the registry were interviewed on short order. The findings were not confirmatory (4).

The storehouse of data from the routine interviews has also been entered on short notice to test etiologic hypotheses. For example, the interviews have included questions on the use of oral contraceptives and answers have permitted evaluation of some of the hypotheses raised about these agents (5).

We have seen that the process of untangling human teratogens is a complex one that requires the energies and cooperation of many people with various backgrounds and approaches. Birth defect surveillance and related epidemiologic approaches can play a role in the continuing effort to find human teratogens and to evaluate suspected teratogens. If you have concerns about an environmental agent being a teratogen and feel that we have data that may be of use in testing this hypothesis, we will be happy to help you in any way that we can.

REFERENCES

1. Janerich, D. T., Piper, J. M. and Glebates, D. M.: Oral contraceptives and congenital limb-reduction defects. N. Engl. J. Med. 291:697, 1974.
2. Record, R. G. and Armstrong, E.: Incidence of congenital limb-reduction deformities. Lancet 1:804, 1975.
3. Center for Disease Control. Epidemiologic Notes and Reports, Spray Adhesives, Birth Defects, and Chromosomal Damage. Morbidity and Mortality Report 22:365–366, 1973.
4. Rachelefsky, G. S., Flynt, J. W., Jr., Ebbin, A. J. and Wilson, M. G.: Possible teratogenicity of tricyclic antidepressants. Lancet 1:838, 1972.
5. Oakley, G. P., Jr., Flynt, J. W., Jr. and Falek, A.: Hormonal pregnancy tests and congenital malformations. Lancet 2:256, 1973.

Latency of Defects*

Edmond A. Murphy, MD

The mechanisms which dictate whether or not a genetic defect becomes
clinically manifest are of considerable intricacy. It is difficult to construct a co-
herent scheme to encompass them and to make sure that none has been overlooked. A number of such mechanisms will be discussed. They are arranged in
order of descending importance of the genetic factors and increasing importance
of the environmental factors. Each of the 5 major topics will be discussed.

GENETIC MECHANISMS

Allelic Interaction

The most obvious reason why a trait is not manifested is that genetically
it behaves as a recessive. This idea is commonplace, though in general it is not
obvious why a trait should be dominant or recessive. Broadly, if there is a good
deal of redundancy in the particular locus and if the mutant gene has no regu-
lator product, or has a product which is enzymatically inert, one may expect
the trait to be recessive. If there is no redundancy of the product at the locus,
or if the product is defective structurally, then one may expect it to be
dominant.

Two-Locus Interaction

Where two or more loci are involved, it becomes necessary to appeal, at least
in principle, to the theory of networks which is encountered in electic circuitry
or traffic engineering. In such systems 3 questions are of cardinal importance:

*Supported by NIH grant 5 P01 6M19489 and a grant from The National Foundation —
March of Dimes.

Birth Defects: Original Article Series, Volume XII, Number 5, pages 7—18
© 1976 The National Foundation

1) Are the various effects (transformed if necessary) additive? Or is their joint effect determined by the minimum component? For instance, we know that the rate at which bone is formed depends on the product of the serum concentration of calcium ion and that of phosphate ion; on a logarithmic scale the effects of these two components would be additive. On the other hand, anemia depends both on the supply of hemoglobin and the supply of blood cells to contain the hemoglobin. Thus the amount of anemia will be determined by whichever is the less.

2) If there are two components in the production of a desired product, are they arranged in parallel or in series? In the two-locus case, an arrangement in series is commonly called an *epistasis*. For instance, if substance H is not produced then the blood group substances A and B cannot be produced; hence, as is well known, the homozygous state of the Bombay blood group gene completely obscures what is going on at the ABO blood group locus (1). Alternatively, the two (or multiple) loci may be acting in parallel and this is what we have called elsewhere a *parastasis* (2). Thus there may be multiple pathways through which some desired product may be produced. There is, for example, evidence of at least two, and perhaps more, loci which produce α hemoglobin chains (3). The difference between epistasis and parastasis is that when the processes lie in series, disruption of *any* one step will prevent the entire process; if they are arranged in parallel, it is only when there is a disruption of *all* pathways that the product will be totally deficient.

3) Is there a cybernetic process involved? In other words, does the deficiency of a product lead to compensatory mechanisms (as anemia stimulates erythropoietin production)? Or stimulate an alternate product (as β thalassemia promotes production of γ chains)? Or does it make one product reflexive (as in hemoglobin H or hemoglobin Barts in which all the chains are of one type)?

Clearly there is quite enough here to give rise to a rather elaborate theory. By way of illustration and for simplicity, I shall confine myself to the two-locus case where the pace is set by a rate-limiting step and there is no feedback. The simplest useful illustration is shown in Figure 1 in which we suppose substance A is converted into substance B through one pathway and then B into C through alternate pathways. The rate of the entire system will depend on the activity of the first step, or the combined activities of the second step, whichever is the less. Four fundamental types are shown. Where the first step is overwhelmingly the more active, the response will depend on the union of the second steps. If the competing steps at the second stage are of about the same magnitude, there will be a graduated response and we have in fact a multifactorial system in the galtonian sense (Fig. 1A). If there are two very unequal pathways and the more important becomes defective, we have such conditions as the pyridoxine-sensitive homocystinuria in which the minor pathway may be excited into activity by pharmacologic doses of the coenzyme (Fig. 1B). There may, of course, be no

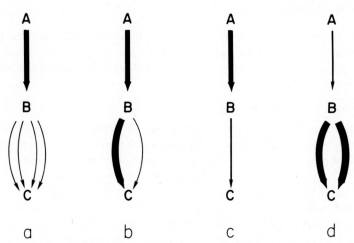

Fig. 1. Four types of simple interactions between loci. A) The rate-limiting step is parastatic. B) The second step consists of an unequal parastasis. C) Both steps are simple, the second being rate-limiting. D) The first step is rate-limiting, the second parastatic.

alternate pathway at all, ie there is no parastasis at the second step; and a defect in the sole pathway results in an inborn error of metabolism (Fig. 1C). It is pertinent to note that inborn errors of metabolism are associated with deficiencies of enzymes of high specificity. Where there are multiple enzymes (as with the alkaline phosphatases), no inborn error of metabolism is known. Finally we have the case in which the pathways from B to C are both very wide and that from A to B very narrow (Fig. 1D). In these circumstances it is only the first step which matters; the enzymes at the second step are presumably detectable biochemically but not recognized as genetic disorders. This presentation is rather oversimplified. A somewhat more quantitative treatment of it is given elsewhere (2).

Multiple Locus Systems

All these ideas can be carried over to systems in which there is interaction between multiple loci except, as might be imagined, that they make it extremely complex — so complex that they are perhaps to be elucidated by biochemical analysis only and mathematics is likely to contribute elegance but no insight. However, one important change in ideas results. The notion of the normal becomes much more ambiguous where multiple loci are involved (4–6). When a multifactorial system exists, it is no longer possible to think in terms of "the good" and "the bad" as we do with abnormal hemoglobins; instead we consider the golden mean between the deficient and the excessive. Of course, the purely quantitative approach to multifactorial traits with no attempt to identify the operation of individual genes makes any real system of classification impossible. A notable exception is the threshold phenomenon.

THRESHOLD PHENOMENA

The presence or absence of abnormality may in no way reflect any true intrinsic genetic process but the relationship of the phenotypic value to some critical value (threshold). Such critical values can be conveniently grouped into real and apparent.

Real Values

A real threshold value means one which does not depend on arbitrary judgment or on some artifact of observation, but which involves some real physical or chemical property of the structure concerned. For instance, the force required to break a bone depends on the qualities of the bone, not on those of the orthopedic surgeon. Of course we know that there are disorders in which bones are excessively fragile, but equally we know that fragility of bone depends on such factors as age. As Edwards pointed out many years ago (7), rupture of a cerebral blood vessel may be a critical single event and yet its origin may be based on blood pressure which is evidently multifactorial. Calculus formation may be a matter of accidental crystallization from a supersaturated solution, a phenomenon which will not occur if the saturation point is not exceeded.

However, we know that real thresholds do exist for traits which — there is persuasive, or even compelling, evidence to believe — have a unilocal origin. I give three such mechanisms.

Environment. Wilson disease can be prevented or (in some degree) reversed by removal of excessive copper from the body. Copper is obviously derived mainly from the environment and it seems reasonable to infer that the delay in onset of this disease is determined by the time required for accumulation of sufficient dietary copper. The same kind of mechanism has been put forward to explain hemochromatosis and there are doubtless many other examples of heavy metal poisoning of this type. It seems reasonable in developing this idea to think in terms of a reservoir with acceptable upper and lower limits. The latency of the defect is presumably determined by the size of the reservoir (Fig. 2). Thus the blood sugar reservoir is a small one, and the effects of deprivation of sugar are promptly made apparent in the neonate. The phenylalanine reservoir is somewhat larger and the ill effects of overloading are somewhat more slowly apparent. In Wilson disease the reservoir for copper is presumably considerably greater. Trouble may arise either because of overflow or underflow. Copper absorption is (in the circumstances) probably too efficient in Wilson disease; it is grossly inefficient in Menkes syndrome (8). In broad terms, what is going on in the reservoir will depend on imbalance between input and output. Either or both of these factors may be environmental or genetic.

Development. The existence of a threshold cannot be detected until it has been challenged. Quite often then, the latency of a trait depends on some process of normal development. For instance, hematometrocolpos is not usually

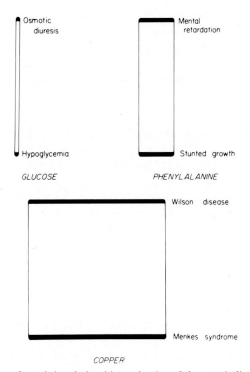

Fig. 2. Penetrance of a trait in relationship to the size of the metabolic pool. For blood sugar, the size of the pool is small. Thus a defect tending to under − or over − fill the pool will rapidly become evident. The pool for copper is large, and disease due to excess or deficiency will only slowly become apparent. Phenylalanine is intermediate.

detected until menstruation has started. The majority of cases of testicular feminization are not discovered until puberty unless suspicion has arisen that it is segregating within the family. Often the Turner syndrome is not detected until puberty. Again, disorders of growth may not be detected or diagnosed at birth but elucidated only by the evolution of the condition. Thus, in deficiency of growth hormone, birthweight is normal but there is a failure to grow adequately. Again, mental deficiency may not be evident at birth, nor deafness nor blindness. Before the age of about 2 years, it would be difficult to determine whether a child is color-blind. There are a great many conditions with delayed onsets (such as Tay-Sachs disease, metachromatic leukodystrophy, Duchenne muscular dystrophy and so forth) which may conceivably be precipitated by the process of development, though they may also be due to other factors which I shall discuss later.

Genetic factors. Even for traits which are known to be under the control of a single genetic locus, the expression of the phenotype may be determined by what are apparently purely random genetic factors. Of these the most commonly invoked mechanism is lyonization. Whether or not there is an analogous process for autosomal traits is a moot point, but where measurements have been made, there is always a greater variance for the heterozygote than there is for either of the homozygous conditions. This finding may be an artifact of scaling, but it may be due to some process of inactivation for the autosomal traits.

Apparent Values

The threshold for the appearance of the disease is commonly dictated not by intrinsic qualities such as we have discussed but rather by what are purely extrinsic, and therefore accidental, qualities. No doubt there are many mechanisms involved; I can think of two.

Arbitrary. Critical levels may be determined either by administrative convenience or by the prejudices of one's peers. Thus, what constitutes mental deficiency may depend on some legal prescription. If it is argued that those who are mentally deficient, and hence do not fully understand the nature of what they are doing, require special protection from the law, presumably the law must draw an arbitrary dividing line at some point. The problem has been discussed at length (6). Such a system has utility; however, it gives rise to certain anomalies. A person under consideration may sometimes meet the criteria of barely acceptable normality and sometimes may not. In effect he is rated mentally deficient on random days of the week. Or again, the Little People of America have decided that nobody who has attained the height of 5 feet or over should be eligible for membership. Such a level is much less ambiguous than an intelligence quotient; nevertheless, it is just as arbitrary. Both height and intelligence (so far as we know) are under the control of multiple loci and the variance which is introduced by multiple segregations will make it a matter of chance whether the child of some particular progeny falls into some arbitrary class or not. These classes do have practical utility, but they do not constitute any mechanism of scientific interest except perhaps sociologic or psychologic.

Diagnostic sensitivity. Whether or not a person be judged abnormal depends on the sensitivity of the methods which are available for diagnosis. In many families supravalvular aortic stenosis does seem to behave as a more or less regular mendelian trait; nevertheless, there is the occasional skipped generation which has somehow to be accounted for. It may, of course, be due to the operation of an epistatic gene or some such mechanism, but it is also possible that a criterion more refined than feeling for a thrill and listening for a diamond-shaped murmur (eg cardiac catheterization or pathologic inspection of the valve) might show a mild degree of aortic stenosis in these "skipped" people.

Another such instance would be dental changes. The dental stigmata of congenital syphilis appear in the second dentition but not the first. If the other signs of the disorder were unobtrusive, the diagnosis might be missed. In time,

of course, the second dentition will appear and an alert diagnostician would recognize the pathognomonic findings. But equally it might be argued that the second dentition could be analyzed at a much earlier age by x-ray examination and the anomalies discovered earlier. Such instances could be multiplied.

FAIL-SAFE MECHANISMS

Considering the intricacy of the vital processes and the capacity for errors to occur, it is indeed remarkable how successfully we survive and remain healthy. Of course much is achieved by a ruthless process of prenatal and perinatal selection, but perils remain throughout life. As we understand it, a very elaborate system of correcting mechanisms is involved. I think of a hierarchy of mechanisms which successively increase the survivorship of the system (Table 1).

There may be no patrimony for the cell — merely the means to extract sustenance from another cell, as in bacterial viruses. Failure occurs if the host cell is wanting. If the functioning of the cell is less constrained and depends on the availability of some substance, such as glucose, then an initial supply of this substance will enable it to survive on its own for a finite period of time. When the source of substrate is exhausted or the metabolites become overwhelming, the system dies; but then the substrate may be replenished by synthesis or by extraction from the ambience. In general, this process will depend on the operation of enzymes which will in time be destroyed. However, those cells with an endowment of RNA or DNA will replenish the supplies of enzyme and, in a perfect environment, would presumably continue to do so indefinitely. In practice, repeated use of RNA or replication of DNA will lead to structural changes which give rise to an inefficient enzyme product. In some systems there exist mechanisms for the excision of defective RNA and its replacement by normal RNA. However, in certain disorders (such as xeroderma pigmentosum) these mechanisms for the repair of RNA are defective; moreover, it seems likely that they break down even in normal cells. In certain specialized tissues (notably the blood cells, the gonads, the liver, the gastrointestinal tract) the operation of the system may be further preserved by having a mechanism for replacing entire cells. Nevertheless, the mechanism for forming new cells may fail, as in aplastic anemias or the indolent wounds which occur after radiation damage. Rarely, a further back-up system is furnished by differentiation to form a new organ. This probably occurs only in fetal life and in neoplasia. Most commonly, failure follows because no such response occurs.

Finally, organ transplantation may furnish a last line of defense, but too often transplantation fails.

Universally, the endowments appear to be more or less genetic, the failures environmental. A possible exception is the iatrogenic endowment; even here it might be argued that it is the inherited intelligence of man which is operating.

TABLE 1. Orders of Complexity of Genetic Fail-Safe Mechanisms

Level	Endowment		Failure	
	Type of Endowment	Example	Cause of Failure	Example
0	Pure parasitism	Bacterial viruses	No host	—
1	Store of sustenance	Most cells	Exhaustion of supplies	Sweet corn
	Absence of waste products		Accumulation of waste	Storage disease
2	Synthesis or extraction of sustenance and the elimination of waste products	Most cells	Exhaustion of enzymes	Erythrocytes
3	DNA and RNA	Most nucleated cells	Destruction of RNA and DNA by injury	Many cells
4	Repairing mechanisms for DNA	Normal skin cells	Failure of repair	Skin cells in xeroderma pigmentosum
5	Replacement from stem cells	Liver, skin, blood, testes, etc	Destruction of stem cells	Aplasia, radiation damage
6	Dedifferentiation	Fetal cells Extracellulary erythropoiesis	Irreversible histone masking	Neurons
7	Iatrogenic repair	Transplants Dental prostheses	Inadequate technique	Defective tissue typing

MULTIPLE HIT MECHANISMS

I have had a special interest in this field because of our work on the theory of platelet survival (9). It is supposed that a cell or an organism has a certain capacity for surmounting insults, each of which takes its toll until the damage done is incompatible with further survival. Admittedly, this process may, to a greater or lesser extent, be obscured by the fail-safe mechanisms which we considered in the last section. Nevertheless, there are certain instances in which a system of reserves, rather than one of repairs, appears clearly to be at work.

I have not so far used the term *incomplete penetrance*; it is commonly regarded as incongruous in modern science. However, it may usefully be applied to the dependence of the manifestation of a gene on a purely environmental challenge. The challenge could be a single hit affair or depend on a series of insults. Let us consider three familiar examples.

Single hit responses. There are certain idiosyncrasies in response to drugs which may give rise to manifestations at the first exposure. One thinks of pseudocholinesterase deficiency, porphyria, glucose-6-phosphatase dehydrogenase deficiency and favism. Had their biochemical bases not been elucidated, one might think of these traits as being mendelian but with incomplete penetrance.

Two-hit responses. Here one thinks predominantly of allergic and sensitizing reactions. One of the best-resolved examples is sensitization to choramphenicol; but it is well-known that sensitivity to poison ivy increases with contact, and in multiple attacks of acute glomerulonephritis, the incubation period between throat infection and the renal manifestations becomes progressively shorter.

Multiple hit phenomena are much more difficult to demonstrate cogently even in the case of circulating cells which can be most easily studied and which have comparatively simple economies. However, I may point out that multiple hit models have at least been put forward for such diverse conditions as dental caries (10), arthritis (11), coronary disease (12), and senescence (13). One of the nice properties of multiple hit models is that when the number of hits becomes large, the distribution of first abnormality (eg death) becomes first a unimodal and positively skewed curve and then comes very close to a normal distribution (9). The ages of onset of many delayed traits tend to follow these patterns; I think particularly of Huntington chorea in which the age of onset conforms very exactly to a normal distribution (14).

LAX TOLERANCES

The intricate pattern of development of a normal structure suggests it is important that certain timing and spatial relationships should be observed to avoid chaos. Thus, the precise timing of sensitivity to certain teratogens has been worked out in considerable detail in animals and occasionally, as in thalidomide sensitivity, also in man. These rigors are probably mitigated by two factors.

There probably are feedback mechanisms which will, within limits, adjust anatomical abnormalities, as illustrated by the effect of weight-bearing on the fine structure of bones which have been broken and set with good alignment but imperfect apposition. In the nature of things it is difficult to study such mechanisms *in utero.* Secondly, in many cases the timing of events may not be quite so critical as is widely supposed. Thus, the ductus arteriosus ordinarily closes at birth, but closure may be delayed a matter of days, weeks, months or even occasionally years. One notes the studies of Fraser and Pashayan (15) and of Kurisu et al (16) on the relationship between facial configuration and palatal clefting in man. Even more precise studies have been done in mice in which a defective gene tending to produce clefting has been studied against various genetic backgrounds (17). It seems that certain types of genome do influence the risk of facial clefting. On the other hand this is nowhere an all-or-none matter: the risk may be bigger in the one group than in the other but is not zero in any or 100% in any. Such studies may be explained by supposing that one is looking at a major gene against an essentially galtonian pattern. But I do not find this explanation very satisfactory when found in the context of inbred stocks of mice. I am inclined to think that there is a certain amount of "slippage," both spacial and temporal, which are consistent with normal development, and that there is a dose-response relationship between slippage and probability of defect. Such a state of affairs, while clinically comforting, would set limits on the resolving power of genetic analysis in this context.

CONCLUSIONS

The topic which I have undertaken to discuss is of considerable complexity. It is comparatively easy to assemble a list of mechanisms such as I have discussed; it is much more difficult to assess the essential merits and to determine whether the list is exhaustive. This discussion has been entirely devoted to classification and illustration. Nothing has been said about how one is to confront this type of problem genetically. We have dealt with few of these areas in some depth elsewhere (2, 18, 19) and all that it is feasible to do here is to give references. Major contributions in these fields have been made by investigators viewing the topic from a mathematic standpoint (20–23). However, mathematic analysis likely will get very far without a steady flow of empiric information. It is not that models of any degree of complexity cannot be developed; the difficulty is in distinguishing among competing plausible hypotheses in view of the paucity of information likely to accrue even from extensive studies. In the past, geneticists have tended to be concerned far too much with goodness-of-fit tests and far too little with the problem of power. There has been a glimmering of a healthy concern with power in the last few years (24), and I would regard this as indispensible for anyone who is making claims about a model, unless he is content to regard it as a purely descriptive device. However, the capacity to distinguish between competing models is in most cases limited; I think we must look to experimental and empiric analysis of the intermediate mechanisms involved if the study of birth defects is ever going to be put on a sensible footing.

REFERENCES

1. Bhende, Y. M., Deshpande, C. K., Bhatia, H. M. et al: A "new" blood group character related to the ABO system. Lancet 1:903, 1952.
2. Murphy, E. A.: Genetic counseling. In Purkis, I. E. and Matthews, U. F. (eds.): "Medicine in the University and Community of the Future." Halifax: Dalhousie University, 1969, pp. 143–150.
3. Kattamis, C. and Lehmann, H.: The genetical interpretation of hemoglobin H disease. Human Hered. 20:156–164, 1970.
4. Murphy, E. A.: A scientific viewpoint on normalcy. Perspect. Biol. Med. 9:333–348, 1966.
5. Murphy, E. A.: The normal and the perils of the sylleptic argument. Perspect. Biol. Med. 15:566–582, 1972.
6. Murphy, E. A.: The normal. Am. J. Epidemiol. 98:403–411, 1973.
7. Edwards, J. H.: The simulation of Mendelism. Acta Genet. (Basel) 10:63, 1960.
8. Danks, D.M., Stevens, B. J., Campbell, P. E. et al: Menkes kinky-hair syndrome. An inherited defect in the intestinal absorption of copper with widespread effects. In Bergsma, D. (ed.): "Medical Genetics Today," Birth Defects: Orig. Art. Ser., vol. X, no. 10. Baltimore: The Johns Hopkins University Press for The National Foundation – March of Dimes, 1974, pp. 132–137.
9. Murphy, E. A. and Francis, M. E.: The estimation of blood platelet survival II. The multiple hit model. Thromb. Diath. Haemorrh. 25:53–80, 1971.
10. Burch, P. R. J. and Jackson, D.: The greying of hair and the loss of permanent teeth considered as an autoimmune theory of aging. J. Gerontol. 21:522, 1966.
11. Burch, P. R. J.: Autoimmunity: Some etiological aspects. Inflammatory polyarthritis and rheumatoid arthritis. Lancet 1:1253, 1963.
12. Murphy, E. A.: Some difficulties in the investigation of genetic factors in coronary disease. Can. Med. Assoc. J. 97:1181–1192, 1967.
13. Beard, R. E.: Appendix: Note on some mathematical mortality models. In Wolsterholme, G. E. W. and O'Connor, M. (eds.): "The Life Span of Animals," Ciba Foundation Colloquia on Aging. Boston: Little Brown and Co., 1959, vol. 5, p. 302.
14. Murphy, E. A.: The rationale of genetic counselling. J. Pediatr. 72:121–130, 1968.
15. Fraser, F. C. and Pashayan, H.: Relation of face shape to susceptibility to congenital cleft lip: A preliminary report. J. Med. Genet. 7:112–117, 1970.
16. Kurisu, K., Niswander, J. D., Johnston, M. C. and Mazaheri, M.: Facial morphology as an indicator of genetic predisposition to cleft lip and palate. Am. J. Hum. Genet. 26: 702–714, 1974.
17. Trasler, D. C.: Pathogenesis of cleft lip and its relation to embryonic face shape in AJ and C57BL mice. Teratology 1:33, 1968.
18. Murphy, E. A. and Bolling, D. R.: Testing of a single locus hypothesis where there is incomplete separation of the phenotypes. Am. J. Hum. Genet. 19:322–334, 1967.
19. Murphy, E. A. and Chase, G. A.: "Principles of Genetic Counseling." Chicago: Year Book Medical Publishers Inc., 1975.

20. Falconer, D. S.: The inheritance of liability to disease with variable age of onset with particular reference to diabetes mellitus. Ann. Hum. Genet. 31:1, 1967.
21. Edwards, J. H.: Familial predisposition in man. Br. Med. Bull. 25:58, 1969.
22. Smith, C.: Recurrence risks for multifactorial inheritance. Am. J. Hum. Genet. 23: 578, 1971.
23. Mendel, N. R. and Elston, R. C.: Multifactorial qualitative traits: Genetic analysis and prediction of recurrence risks. Biometrics 30:41, 1974.
24. Chung, C. S., Ching, G. H. S. and Morton, N. E.: A genetic study of cleft lip and palate in Hawaii. II. Complex segregation analysis and genetic risks. Am. J. Hum. Genet. 26: 177–188, 1974.

Radiation-Induced Malformations in Man*

Timothy Merz, PhD

The study of developing biologic systems disturbed by radiation has produced a literature perhaps as large and as extensive as any in radiation biology. Practically all of the major groups of vertebrates and some of the minor groups have been studied to determine the effects of radiation on their embryos and fetuses. Even man, a somewhat unusual experimental animal, has been included in a list consisting of monkeys, cows, dogs, sheep, goats, pigs, rabbits, guinea pigs, hamsters, rats, mice, chickens, amphibians and fish. Most of the experiments, regardless of the material, have produced results that are consistent and reproducible from animal to animal, and practically all radiation-induced defects have been elicited in more than one organism.

Man, as a creature of evolution, has evolved in the presence of a background of what might be called natural radiation amounting to approximately 0.1 rem/year, as have most of the other organisms that share our time and space. One would expect, then, according to the rules of evolution, that there would be a number of rather sophisticated enzymatic repair processes to take care of whatever damage ensues from environmental radiation. Indeed, such processes can be demonstrated in bacteria and man alike, and involve such enzymes as endonucleases, exonucleases, ligases and polymerases, all designed to eliminate the damage and replace it with normal, functional, genetic material.

Since radiation is only one of a number of agents in our environment that perturbs biologic systems, it is not surprising to note that there are essentially no known effects induced by radiation that are not produced by other agents and, in fact, already exist as spontaneous phenomena. The change that is seen in the presence of radiation is a change in frequency. The number, not the kinds, of things that happen are altered. Another important point to consider is the concept of thresholds. Experimentally, there is apparently no threshold beneath which a dose of radiation causes no damage. Operationally, the fact is probably not significant since it might take an extraordinarily large population to show

*The information necessary for a complete understanding of this field can be found stated simply in *Radiobiology for the Radiologist* by Eric J. Hall (1), more completely in the BEIR Report (Biological Effects of Ionizing Radiation) (2), AEC Symposium No. 17 (3) and in the voluminous works of Roberts Rugh (4).

Birth Defects: Original Article Series, Volume XII, Number 5, pages 19–22
© 1976 The National Foundation

statistically significant damage done by very small, well-separated doses. Radiation-induced changes are not simply the consequence of damage, but the expression of a series of processes that start with damage which is then altered by repair mechanisms and replacement of damaged cells by undamaged cells. The final consequence is the expression of the residual damage. In other words, there are 2 ways in which initial damage to a cell can go unrealized. The first way is to repair the damage. The second way is for the cell to operate on the small amount of genetic material that remains undamaged.

Malformations are the consequences of unrepaired damage. Generally, "radiation-induced malformations in man" is a phrase used to describe immediate obvious defects in the fetus as an abortus, at birth, or shortly after birth. At the very least, it is considered a structural phenomenon dealing with organs and tissues. It might, however, be more accurate to look at radiation-induced malformations as any induced alteration which will lead to altered or nonfunctional structures. Such an approach would provide a more useful way of addressing the problem of radiation effects on developing organisms since it would allow the inclusion of such events as karyotypic changes and also the processes which precede the clinical manifestations of radiation-induced leukemia in children.

The range of defects that one observes as a result of radiation insult extends from alterations of functions, congenital anomalies, stunting, and the shortening of life expectancy to immediate lethality. Regardless of the dose, the kind of abnormal progeny produced is dependent upon the time of gestation at which the embryo is irradiated. Less than 10 rads of x rays is a critical dose for precleavage stages in a mouse. Later stages, in utero, can withstand almost 600 rads and survive. Part of the reason for the early stage sensitivity of embryos resides in the number of cells that make up the critical target. Unrepaired injury of a few cells in the 2- to 4-cell stage will result in death and reabsorption, while at a later time that loss would not even be noticeable since all of the cells that are blastomeres have the potential to produce normal offspring and can do so up to gastrulation. Early on, then, it is either death or normality which are the 2 responses to moderate radiation.

The onset of differentiation produces an entirely different response spectrum. At this time congenital anomaly production is relatively reproducible if irradiation occurs in each experiment at the same gestation time. The transition from primitive organ cell type to differentiated organ cell type is the period during which the cells appear the most radiosensitive, as assayed by the production of anomalies of that organ. It must be at this time that cells are the least able to repair the damage done to their genetic material, their information center, and it is perhaps at this time that other surviving cells can least substitute for those which are rendered nonfunctional by radiation damage. At the end of organogenesis, most congenital anomalies, excepting a few very important ones, are no longer produced. The general responses at these times are those which result from cell loss and are most often seen as stunting. This stunting can be observed as a diminution of a particular organ or of the entire animal.

The transition between the effect of small doses and the effect of large doses can best be described as a change in frequency. The relationship between the frequency of an event after a large dose of radiation and the frequency of that event after a small dose of radiation is not linear and may well vary by as much as the square of the dose (exponential). That relationship is also dependent upon how the dose of radiation is delivered; as a single dose, fractionated into small doses, or as a constant low chronic exposure.

The mechanisms involved in radiation-induced developmental anomalies most certainly involve changes in genetic material: gene changes, chromosomal changes, karyotypic changes and perhaps virus insertions. Gene changes, resulting from an altered code (base pair changes, deleted bases) are almost always recessive and deleterious. Most gene changes, therefore, do not appear in the first generation. Chromosomal changes, however, immediately become obvious as chromosome rearrangements or as karyotypic changes. There are some obvious and interesting correlations between chromosome alterations and the production of anomalies. Radiation effects are due largely to the production of active radicals, mostly OH radicals, and their combination with biologic material. Chromosomes are considered to be single Watson-Crick helices and most of the recognizable alterations and their structure come from breaks in one or both strands in the helix. Various simple alterations in structure can be formed by either the nonrepair or misrepair of lesions in DNA strands. Karyotypic changes are the result of associations between chromosomes and exchanges (translocations) between these 2 chromosomes or nondisjunction during division. These phenomena are both increased after radiation exposure and may account for the increased number of trisomic offspring produced by irradiated parents. Chromosomal changes seem to be related to cell death as well as the production of anomalies, and the amount of damage is related not only to the size of the dose, but also to the way it is delivered, ie as a chronic or a fractionated dose vs a single acute exposure. The small fractionated doses and the chronic doses tend to allow repair to go on during the production of damage. As a result of repair, there is less interaction between damaged sequences, and lower frequencies of aberrations are observed at the termination of exposure.

The most significant hazards to mankind are probably those which are related to very low doses fractionated over a long period of time: increases in background rates in radiation, increases in clinical diagnostic procedures, increases in the use of nuclear medical techniques. The really significant exposures to human beings in large numbers are not likely the results of nuclear holocaust, nuclear power plants and radiation therapy. Low dose effects may be of greater significance.

Although the subject is controversial, some obstetricians, acting on the present state of knowledge and ignorance about radiation-induced alterations in fetuses, may suggest abortions after fetal exposure in the range of 1 to 10 rads, especially when subjected to this dose in the first 2 trimesters. But what about doses significantly less than that, such as the effect of doses in the range of 0.5 rad on the fetus and in cases where the fetus is a female, on the fetus' ova?

There have been several studies of the production of tumors by low doses of radiation to the fetus, the most well-known being those of Stewart and Kneale (5), which has been entitled "The Oxford Study" and which shows, according to these authors, a great increase in the production of leukemia in individuals irradiated in utero. Another kind of study has been done by Meyer and coworkers (6). That study involves the fetus irradiated in utero and, consequently, the fetus' ova. The ova are of great interest because, although the fetus has reached its minimal response to radiation at the end of a trimester, the ova are theoretically in their most sensitive stage. The study by Meyer and co-workers is still in its preliminary stages but shows some rather interesting results, including altered fertility patterns in irradiated fetuses. Another study of interest is the evidence, soon to be reported by Mullenix et al (7), that the behavior of rats radiated at low doses during development is altered in a reproducible and quantitative fashion.

In conclusion, one should realize the dangers inherent in unnecessary irradiation, especially at low dose levels when the dangers are not so obvious, and instead of viewing all increases in personal and population exposures as trivial, we should view everything in terms of a risk vs benefit situation.

REFERENCES

1. Hall, Eric J.: "Radiobiology for the Radiologists." Hagerstown: Harper & Row, 1973.
2. BEIR Report: "The Effects on Population of Exposure to Low Levels of Ionizing Radiation." National Academy of Science, National Research Council, Washington, D.C., November 1972.
3. AEC Symposium No. 17.
4. Rugh, R.: Effects of ionizing radiations, including radiosotopes, on the placenta and embryo. In Bergsma, D. (ed.): "Symposium on the Placenta," Birth Defects: Orig. Art. Ser., vol. 1, no. 1, White Plains: The National Foundation – March of Dimes, 1965, p. 64.
5. Stewart, A.: "An Epidemiologist Takes a Look at Radiation Risks." DHEW Publication No. (FDA) 73-8024, BRH/DBE 73-2. January 1973.
6. Meyer, M., Diamond, E. L. and Merz, T.: Sex ratio of children born to mothers who had been exposed to x-ray in utero. Johns Hopkins Med. J. 123:123–127, 1968.
7. Mullenix, P., Norton, S. and Culver, B.: Locomotor damage in rats after x-irradiation in utero. Exp. Neurol., 1975. (In press.)

Immediate and Long-Range Effects of Maternal Viral Infection in Pregnancy*

Janet B. Hardy, MD, CM

INTRODUCTION

Interesting parallels exist between conditions which are of genetic origin and those which arise as the result of maternal-fetal infection during pregnancy. Both may be so severe as to result in fetal death or later, infant death; both may have characteristic malformations and/or other manifestations which involve one organ system or multiple systems; and both may be floridly manifest at birth or have a long latent period before the expression of signs and symptoms. However, even though there may be certain similarities in clinical manifestations, the underlying etiologic mechanisms are different and an understanding of the mechanisms is essential for prevention.

A number of agents capable of producing devastating fetal infections are listed in Table 1. Toxoplasma and T. pallidum, while not viruses, are included in this discussion because the congenital infections they produce are in many ways similar to those produced by rubella, cytomegalovirus and herpes simplex infections and can often only be differentiated by specific laboratory tests. They all can produce chronic infection with a spectrum of abnormalities from the lethal to clinically inapparent infection; however, the risk of significant morbidity is generally high.

Because of small numbers of cases or lack of precise diagnostic information, the risk of abnormality cannot be assessed in the following maternal infections which have caused recognized fetal abnormality: measles, mumps, varicella, variola, vaccinia, hepatitis, poliomyelitis, virus pneumonia, Coxsackie B and influenza.

*This study was supported by NIH, National Institute of Neurological and Communication Disorders and Stroke, grant NB 02371, contracts PH 43 68 12 and PH 43 68 710.

Birth Defects: Original Article Series, Volume XII, Number 5, pages 23–31
© 1976 The National Foundation

TABLE 1. Maternal Infections With High Risk for Fetus

Agent	Death	Malformation	Increased Risk Low Birthweight	Chronic Infection
Rubella virus	+	+	+	+
CMV	+	+	+	+
Herpes simplex	+	+	+	?
Toxoplasma	+	+	+	?
T. Pallidum	+	+	+	+

Finally, there is a group of agents, such as the ECHO and adenoviruses and some members of the Coxsackie virus family, which do not seem to represent a risk to the fetus. The fetal consequences of maternal viral infection in pregnancy were reviewed by Hardy in 1973 (1); these and other agents were discussed.

While congenital syphilitic infections had been recognized for many years, it was Sir Norman Gregg (2) who called attention to the teratogenic potential of maternal viral infection. His description of the triad of congenital cataract, heart disease and deafness following maternal rubella in early pregnancy is a medical classic. His work, and that of others, following the great rubella epidemic in Australia in 1939–40 provided a broad picture of the devastating effects of fetal infection by this otherwise rather benign agent.

Technologic advances in immunology and virology during the past 15 years have provided the tools for the specific diagnostic and epidemiologic study of the pathogenesis and consequences of many maternal-fetal infections. The isolation of rubella virus in 1962 is of particular importance (3, 4). Viral isolation made possible: 1) the development of diagnostic serologic tests necessary for definitive study and 2) a vaccine for the active immunization of susceptible individuals.

The Johns Hopkins Child Development Study (CDS) was a part of the Collaborative Perinatal Project (CPP) of the National Institute of Neurological and Communication Disorders and Stroke (NINCDS). The CPP was a prospective, multidisciplinary, longitudinal investigation of the cause of congenital malformation and neurologic, neurosensory and intellectual defects of infants and children. To achieve this goal detailed information was collected during the mother's pregnancy, labor and delivery, and each child was followed with a planned protocol of neurologic and behavioral examinations until the age of 8 years (5). The Johns Hopkins Study population included approximately 5000 pregnancies terminating between late 1958 and August 1965; it has been described elsewhere (6).

The routine collection of clinical information and serum samples during pregnancy from each woman enrolled provided for an investigation of the role of maternal infection during pregnancy in relation to both the immediate and long-range outcome for the child. There was a widespread epidemic of Asian influenza just as the Johns Hopkins Study began (7), and the great rubella epidemic of 1963–64 was just over as the last study baby was born in 1965.

The influenza studies provided suggestive evidence of an increase in the rate of congenital malformations among the babies of mothers having serologic evidence of influenza during the first trimester.

The 1370 women delivering infants during 1964 and early 1965 were included in a special ancillary investigation of congenital rubella. Among this group of pregnancies, 130 were complicated by maternal infection with rubella virus — a rate of almost 10%. Approximately 50 women had clinical illness including rashes; in the remainder the diagnosis was based on laboratory findings as the mother's infection was inapparent. The ongoing prospective study offered an unusual opportunity to identify the florid manifestations of congenital infection resulting from maternal infection during the first 3 months of pregnancy and, more importantly, the subtle but nonetheless handicapping defects resulting from maternal infection during the second trimester of pregnancy.

Two hundred infants with congenital rubella, identified in the wards and clinics of the Johns Hopkins Hospital were added to the group of CDS children. The entire group of approximately 300 children made up the population of the Johns Hopkins Rubella Study. The laboratory studies, for the most part, have been carried out under the direction of Dr. John L. Sever, in the Perinatal Infectious Disease Laboratory, NINCDS. The clinical studies have been carried out at Johns Hopkins and most of the children are still being followed 10 years later.

PATHOGENESIS OF MATERNAL-FETAL INFECTION

We now turn to a discussion of the pathogenesis of maternal-fetal infection using rubella virus as a model. A diagrammatic representation of the course of events is provided in Figure 1. A susceptible pregnant woman is exposed and infected. She develops an acute infection, which may or may not be accompanied by clinical signs and symptoms. Toward the end of the incubation period and during the acute stage, she sheds virus from the throat. A viremia occurs, which lasts from a few days to 4 weeks, during which time the placenta and the fetus may be infected by way of placental circulation. Töndury (8) has shown that the earliest fetal lesions are found in the lining of the blood vessels and result from infected emboli from the placenta.

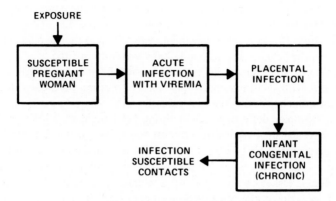

Fig. 1. Pathogenesis of congenital infection.

Fetal tissues are extraordinarily susceptible to infection with both rubella and CMV. The virus replicates in the protected intracellular environment and as the cells divide the virus is passed on in the daughter cells (9). Thus, infected clones arise, made up of cells which are smaller, grow more slowly and die more readily than normal cells. These have important clinical implications and, depending on fetal age at the time of infection, may lead to actual malformation and/or small size. The pathogenic processes responsible for the many manifestations of congenital infection depend upon 2 mechanisms. 1) There may be a direct teratogenic effect resulting from interference with normal patterns of development. Critical cells may be lost at a time of very rapid cell division and differentiation; organogenesis fails to proceed normally and malformations result. 2) A chronic infection may persist, producing a variety of manifestations which may be superimposed on the teratogenic process described above.

The extent and long-range consequences of the fetal infection depend upon a number of variables, the most important of which is the gestational age of the fetus at the time of infection. Table 2 demonstrates the relationship between fetal outcome and gestational age (in weeks) at the time of appearance of the mother's rubella rash. The fetal infections were all documented by the isolation of rubella virus and/or the presence of specific serum antibody after 6 months of age. There is a strong relationship and gestational age plays an important role in determining the outcome. Of the 102 cases studied, 5 mothers had rubella preconception. Four of these women had the disease in the month prior to conception, but the fifth had her disease 3 months before. Of the 5 fetuses, only 1 multiply handicapped child survived. A high risk of abnormality was observed following rubella during the early weeks of pregnancy. While the risk diminished progressively as gestational age advanced, it was of concern until midpregnancy.

These data are supported by those shown in Table 3 which summarizes the findings in the 24 cases in the CDS where the mother had documented rubella in the second trimester of pregnancy. Among these patients, there were 2 fetal losses — 1 following rubella at 14 weeks and 1 at 20 weeks (rubella virus was recovered in both instances). Among the 22 liveborn infants, only 7 have had no evidence of congenital infection. Ten of the 15 infected had hearing and/or specific language disorders, 4 were retarded and 4 had cardiac murmurs (2 of which were due to peripheral pulmonic stenosis).

The relationships between gestational age at infection with CMV, toxoplasmosis or syphilis and fetal outcome are less clear. Since CMV is a chronic,

TABLE 2. The Johns Hopkins Rubella Study.
Gestational Age at Time of Maternal Rubella in Relation to Fetal Outcome

Gestation—		Survived					Died	
		Defects		Mild	No Defects			
completed weeks	No.	Severe	Moderate		? Normal	Normal	Fetal	Later
Preconception	5	1	–	–	–	–	2	2
0–4	23	11	6	–	–	–	2	4
5–8	28	7	9	7	1	1	2	1
9–12	14	3	3	7	1	–	–	–
13–16	10	2	3	1	2	–	1	1
17–20	7	1	–	1	1	3	1	–
21–30	11	1	2	2	2	4	–	–
31–45	4	–	–	1	2	1	–	–
TOTALS	102	26	23	19	9	9	8	8

TABLE 3. The Johns Hopkins Collaborative Perinatal Project.
Relationship Between Maternal Rubella With Rash* After First Trimester and Fetal Outcome

Pregnancy Outcome	Number in Group	Specific Laboratory Findings in Child		
		Viral Isolation	Positive Serology	Elevated Cord IgM†
Fetal loss	2	2/2	Not done	Not done
Liveborn	22	3/19	10/22	9/21
Normal	7	0	0	1/6
Suspect or abnormal‡	15	3/12	10/15	8/15

*Serologic findings were compatible with the diagnosis in each case and 12 women had a clear-cut seroconversion. Surviving children were followed 4 years.
†In 2 children the level was 19 mg%; in the remainder it was in excess of 20 mg%.
‡While details of abnormalities have been reported elsewhere (10), 10 of 15 children had problems in communication.

largely inapparent infection, its onset is difficult to "date." There is some suggestion that toxoplasmosis late in pregnancy may be of more serious consequence than earlier infection. Congenital herpes infection, particularly with type II (the genital strain), tends to occur most often in relation to vaginal delivery and is then an intrapartum infection.

CONGENITAL INFECTION AS SOURCE OF CONTAGION

Because congenital infections are chronic, persistent infections, no great difficulty is encountered in recovering organisms from fetal tissues or body fluids. Numerous investigators have recovered rubella virus from infants for many months after birth. These babies have been documented as infectious (11).

CLINICAL MANIFESTATIONS OF CONGENITAL INFECTION

There have been numerous descriptions of the overt manifestations of congenital infection during the newborn period (1).

The early manifestations may include: 1) cataracts, glaucoma, and structural malformations of heart, eye, ear, palate, dentition, brain and kidney; and 2) widespread evidence of chronic infection of brain, meninges, lung, liver, bone marrow and other sites, that may lead to the inadequate function and small body size which are encountered with both early and later infection. These are well known but the more subtle problems are less well recognized — some are only just beginning to surface as the children grow older.

LONG-RANGE OUTCOME

Intellectual Outcome

Table 4 summarizes the intellectual outcome, at 4—5 years of age, of 171 children with congenital rubella. These rather discouraging results have not improved as the children have grown older. Only 12%, instead of an expected 25% of the children, have IQ scores of 110 or above, and 29%, instead of an expected 3%, scored below 70 IQ. An additional cause for concern is the fact that as the children in the normal IQ range (90 and above) have gone along in school, many are experiencing subtle learning disorders and progress is difficult. Miriam Hardy (12), in her follow-up of hearing-impaired children with rubella, reports that only 25% of those born between 1943 and 1963 and followed into the school years had no defect other than hearing in processing information; there were many problems in perception, memory, sequencing and recall which interfered with normal functioning.

TABLE 4. Intellectual Status of 171 Children With Congenital Rubella (At 4–5 Years)

IQ Score	Number	%	Abnormalities				
			Hearing	Visual	Cardiac	CNS	None
≥ 110 (above average)	20	12	15	1	7	3	4
90–109 (average)	49	29	21	5	15	7	6
70–89 (low normal)	53	30	29	11	22	19	1
≤ 70 (defective)	49	29	28	21	23	30	1
Total number	171	–	93	38	67	59	12
Percent	–	100	54	22	39	34	7

Auditory Status

Approximately 70% of the children in the Johns Hopkins Rubella Study have hearing and/or specific language defects. Most have had numerous audiologic evaluations during follow-up. In about 25% of the children, the hearing loss has been unstable. In a very small number, there seems to have been some improvement in auditory function. In several children between the ages of 5 and 12 years, there has been a sudden, virtually complete loss of residual hearing, and in the rest a progressive decline in function to profound deafness.

Body Size

Many of the congenitally infected rubella infants are growth retarded at birth. They tend to remain small (below 10% for race and sex) through the early years of childhood, particularly in the dimensions of height and circumference. They are expected to remain small. Others, while not small initially, have failed to grow adequately, falling significantly below normal during the first 6–12 months.

Chronic Panencephalitis

Two recent reports describe a chronic, progressive panencephalitis, with late onset, beginning in the second decade, in children whose rubella problems had seemed generally stable to that time. Rubella virus was isolated from the brain in 1 of 4 cases reported (13, 14).

OTHER CONGENITAL INFECTIONS – CYTOMEGALOVIRUS, HERPES

These agents all have the capability to produce congenital infections which range from severe systemic involvement of many organ systems to inapparent infection without obvious abnormality. Specific laboratory studies may be necessary to identify the infection and the nature of the agent.

CMV is widespread in its distribution. The prevalence of CF antibody, which varies with social class and geographic location, is as high as 81% in some areas (15). Initial reports suggested that congenital infections were severe and generally fatal. However, more recent work indicates that, although most congenital infections are mild or inapparent, the disease may be an important cause of mental retardation, congenital deafness and learning difficulties (16, 17). Virus excretion from congenitally infected children may continue for many years, with the infected children being a source of contagion for susceptible contacts. The virus is spread by close personal contact. Since it is present in seminal fluid and persists in cervical secretions, it may be among the sexually transmitted diseases.

SUMMARY

A brief description of the pathogenesis and immediate and long-range outcome following certain maternal viral infections in pregnancy has been presented. As evidence has accumulated from the follow-up of infants with specific congenital infections, it is clear that the consequences can be disastrous but not just for the obviously affected neonate – the normal-appearing infant at birth may have problems which appear much later. So-called "silent" congenital infections of the central nervous system may be a more frequent cause of mental retardation and specific learning disorders than we realize.

REFERENCES

1. Hardy, J. B.: Fetal consequences of maternal viral infections in pregnancy. Arch. Otolaryngol. 98:218–227, 1973.
2. Gregg, N. M.: Congenital cataract following German measles in the mother. Trans. Ophthalmol. Soc. Aust. 3:35–46, 1941.
3. Weller, T. H. and Neva, F. A.: Propagation in tissue culture of cytopathic agents from patients with rubella-like illness. Proc. Soc. Exp. Biol. Med. 111:215–225, 1962.
4. Parkman, P. D., Buescher, E. L. and Artenstein, M. S.: Recovery of rubella virus from army recruits. Proc. Soc. Exp. Biol. Med. 111:225–230, 1962.
5. Niswander, K. R. and Gordon, M.: The Collaborative Perinatal Study of the National Institute of Neurological and Communication Disorders and Stroke. In "The Women and Their Pregnancies." Philadelphia: W. B. Saunders Co., 1972.
6. Hardy, J. B.: The Johns Hopkins Collaborative Perinatal Project: Descriptive background. Proceedings of a symposium. Johns Hopkins Med. J. 128:2–7, 1971.

7. Hardy, J. B., Azarowicz, E. N., Mannini, A. et al: The effect of Asian influenza on the outcome of pregnancy, Baltimore 1957–58. Am. J. Public Health 51:1182–1188, 1961.
8. Töndury, G.: On the infection path and pathogenesis of virus caused damage in the human embryo. Bull. Schweiz. Akad. Med. Wiss. 20:379–396, 1964.
9. Rawls, W. E. and Melnick, J. L.: Rubella virus carrier cultures derived from congenitally infected infants. J. Exp. Med. Sci. 123:795–816, 1966.
10. Hardy, J. B., McCracken, G. H., Gilkeson, M. R. and Sever, J. L.: Adverse fetal outcome following maternal rubella after the first trimester of pregnancy. JAMA 207: 2414–2420, 1969.
11. Hardy, J. B., Monif, G. R. G., Medearis, D. N. and Sever, J. L.: Postnatal transmission of rubella virus to nurses. (Letter to editor.) JAMA 191:1034, 1965.
12. Hardy, M. P., Haskins, H. L., Hardy, W. G. and Shimizu, H.: Rubella: Audiologic evaluation and follow-up. Arch. Otolaryngol. 98:237–245, 1973.
13. Townsend, J. J., Baringer, R. J., Wolinsky, J. S. et al: Progressive rubella panencephalitis: Late onset after congenital rubella. N. Engl. J. Med. 292:990–993, 1975.
14. Weil, M. L., Itabashi, H. H., Cremer, N. E. et al: Chronic progessive panencephalitis due to rubella virus simulating SSPE. N. Engl. J. Med. 292:994, 1975.
15. Birnbaum, G., Lynch, J. I., Margileth, A. M. et al: Cytomegalovirus infections in newborn infants. J. Pediatr. 75:789–795, 1969.
16. Melish, M. E. and Hanshaw, J. B.: Congenital cytomegalovirus infection. Am. J. Dis. Child. 26:190–194, 1973.
17. Hanshaw, J. B.: "Silent" CMV, toxoplasmosis may be a retardation cause. Presented at a symposium on infections of the fetus and the newborn infant. New York University Medical Center, 1975.

Embryopathy Associated With Oral Anticoagulant Therapy*

Judith G. Hall, MD

In 1966 DiSaia (1) reported a child with multiple congenital anomalies, born to a mother who had been on Coumadin (sodium warfarin) during most of her pregnancy. Since then, another 9 similarly affected cases (2–11), whose mothers had taken Coumadin through most of their pregnancies, have been recognized. In addition, another infant was recently reported whose mother was on phenindione (a different oral anticoagulant, but closely related to Coumadin and also a vitamin K antagonist) (9). These 11 cases are summarized in Table 1.

The most consistent clinical feature has been a hypoplastic nose (Fig. 1). The nasal cartilage is underdeveloped, leading to a small upturned nose, often flattened and sunken into the face. The nares are usually quite small with narrowed passages. Choanal stenosis has been diagnosed in 2 cases (2, 7). Marked respiratory difficulty was present in 5 cases (2, 7, 9, 11), requiring an oropharyngeal airway for as long as a month. With time, the noses usually improved somewhat cosmetically but remained relatively small (Fig. 2).

The second most common feature has been bone abnormalities, the most striking of which is stippling (Fig. 3), particularly of the vertebral column, most dramatically the sacrum area. However, the long bone epiphyses, the calcaneus, terminal phalanges and even nasal bones have been involved. Broad hands have been described in 5 cases (6, 8–10). The skull is said to be unusual in 3 cases (1, 5, 9), with prominent occiput and extra fontanels. With growth and ossification, the stippled areas have been incorporated into normal bone and asymmetric growth has not been reported.

Two of the cases (5, 6), in addition to hypoplastic noses and stippled bones, were severely abnormal. One had extremely short limbs, a cataract and died shortly after birth (6). The other had an occipital meningomyelocele, hydro-

*Supported by NIH grants P01 GM15253–08 and 5S01 RRO5655–07, and The National Foundation — March of Dimes grants CA–90 and 6–75–174.

Birth Defects: Original Article Series, Volume XII, Number 5, pages 33–37

TABLE 1. Embryopathy Associated With Oral Anticoagulant Therapy

Hypoplastic nose		11/11
Bone abnormalities		
(stippling, short hands, skull)		10/11
Eye abnormalities		5/11
optic atrophy	2	
cataract	1	
microophthalmia	1	
large prominent eyes with under-developed eyelids at 15-week abortion	1	
Developmental delay		4/9
severe delay – gross malformations	1	
mild-moderate delay		
with optic atrophy	2	
with seizures and deafness	1	
development within normal limits	5	
Intrauterine growth retardation		3/11
Severe malformations		2/11

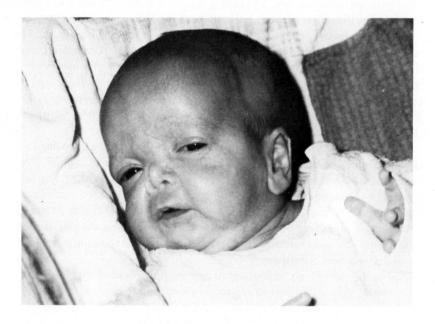

Fig. 1. Patient at age 3 weeks. Note hypoplastic nose flattened into face with small alae nasi.

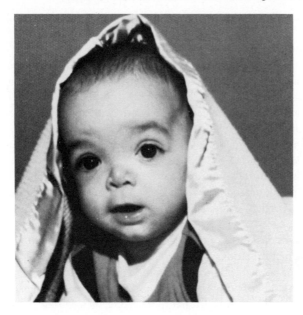

Fig. 2. Patient at age 10 months. Note hypertelorism and relative improvement of nose size.

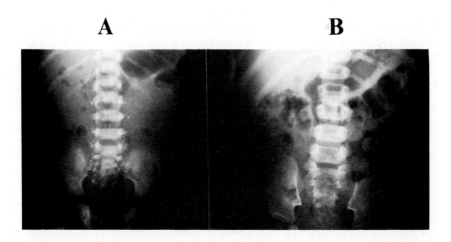

Fig. 3. A) Note stippling along vertebral column, particularly in the sacral area. B) The stippling has been incorporated into normal bone with some mild residual abnormality in sacral area.

cephalus, persistent truncus arteriosus, bilateral microophthalmia and blindness (5).

Mild-to-moderate developmental delay has been described in 4 of the patients — each with an additional problem (Table 1). Five patients were reported as developmentally normal for age, but 2 are too young to really determine. Thus, of known cases, about one third appear to have normal intelligence, one third are retarded and the rest have been severely affected, aborted, died or are too young to know what to expect.

Eye abnormalities were described in the more severely affected cases (1, 5, 6, 8, 9) (Table 1).

The association of these unusual malformations with maternal anticoagulation by a vitamin K antagonist, which is also a very rare occurrence during the first trimester, strongly suggests that these drugs have a teratogenic effect in some embryos.

Multiple cases of the use of Coumadin during the second and third trimester without these malformations have been reported. In addition, there have been many reports of normal children when mothers have taken an oral anticoagulant drug during the first trimester. The exact incidence of these malformations is not clear. Initially, we thought that these problems were associated with the ingestion by the mother of only warfarin derivatives, but it would appear that maternal ingestion of at least one other oral anticoagulant can produce the same kind of problems.

One of the very interesting questions which these cases raise is whether the teratogenic effect is brought about by decreased clotting factors and bleeding or by a vitamin K antagonist effect. It was tempting to imagine bleeding occurring into embryonic tissues with subsequent scarring and secondary calcification. If the anticoagulation is responsible for these problems, then other anticoagulating drugs which can cross the placenta, such as aspirin, would be of concern. However, clotting factors, as we measure them in adults, are absent during the first trimester (12), and cartilage, which appears to be the most consistently affected tissue, has a poorly developed blood supply. So, instead, perhaps the absence of vitamin K in growing tissue leads to the problems. Clearly, there is much to be learned from these cases about normal embryogenesis of cartilage and the role of vitamin K in embryogenesis.

Finally, these cases were all thought to be Conradi syndrome initially, because of the stippling. There are many causes of stippling, but the degree and distribution seen in these cases is most consistent with Conradi syndrome. In 1971 Spranger et al (13) made a very helpful distinction between the autosomal recessive rhizomelic type of Conradi syndrome and the autosomal dominant asymmetric Conradi-Hünermann type. These cases would seem to represent a third category of an environmentally produced phenocopy. There is some suggestion from the literature that other environmental agents, in addition to the vitamin K antagonists, may give similar phenocopies (7).

In summary, the use of oral anticoagulants, in particular the vitamin K antagonist, when taken by a mother during the first trimester — probably the second month — can be associated with malformations in the child. Mothers needing these drugs or who have been taking them should be informed of their risk and offered alternative therapy or termination of the pregnancy.

ACKNOWLEDGMENTS

The author wishes to acknowledge the cooperation of Drs. William Shaul, Elizabeth Newkom and Thomas Shepard, and the technical and secretarial assistance of Ellen Helseth.

REFERENCES

1. DiSaia, P. J.: Pregnancy and delivery of a patient with a Starr-Edwards mitral valve prosthesis. Obstet. Gynecol. 28:469–472, 1966.
2. Kerber, I. J., Warr, O. S. and Richardson, C.: Pregnancy in a patient with a prosthetic mitral valve. JAMA 203:223–225, 1968.
3. Holmes, B., Moser, H. W., Halldorsson, S. et al: "Mental Retardation: An Atlas of Disease with Associated Physical Abnormalities." New York: Macmillan Company, 1972.
4. Wilroy, R. S. and Summitt, R. L.: Personal communication.
5. Tejani, N.: Anticoagulant therapy with cardiac valve prosthesis during pregnancy. Obstet. Gynecol. 42:785–793, 1973.
6. Becker, M. H., Genieser, N. B., Finegold, M. et al: Chondrodysplasia punctata. Is maternal warfarin therapy a factor? Am. J. Dis. Child. 129:356–359, 1975.
7. Shaul, W. L., Emery, H. and Hall, J. G.: Chrondrodysplasia punctata and maternal warfarin use during pregnancy. Am. J. Dis. Child. 129:360–362, 1975.
8. Burdi, A. R.: Personal communication. To be published.
9. Pettifor, J. M. and Benson, R.: Congenital malformations associated with the administration of oral anticoagulants during pregnancy. J. Pediatr. 86:459–462, 1975.
10. Holmes, L.: Personal communication. To be published.
11. Pauli, R. M.: Personal communication. (Accepted for publication, J. Pediatr.)
12. Bleyer, W. A., Hakami, N. and Shepard, T. H.: The development of hemostasis in the human fetus and new born infant. J. Pediatr. 79:838, 1971.
13. Spranger, J. W., Opitz, J. M. and Bidder, U.: Heterogeneity of chondrodysplasia punctata. Humangenetik. 11:190, 1971.

Cortical Cataracts Following Total Parenteral Nutrition*

J. Denis Catalano, MD **and James A. Monteleone,** MD

Recent advances in the care of ill neonates have introduced many innova-
tions into the medical regimen applied in the intensive care nursery. Among
these are bilirubin lights, assisted pulmonary ventilation, radiant heat incubators
and total parenteral nutrition (TPN). These techniques have not been used long
enough for all possible complications to be detected. After observing bilateral
cataracts in a patient receiving TPN, we performed detailed ophthalmologic
examinations on every patient admitted to the intensive care nursery at
Cardinal Glennon Memorial Hospital.

METHODS

Eighty-two consecutive patients admitted to the intensive care nursery between February 1 and April 1, 1975, were examined ophthalmologically.
Ophthalmologic examination included indirect ophthalmoscopy, retinoscopy
and slit-lamp biomicroscopy. The examination was carried out at the time of
admission and was repeated periodically, depending on the patient's diagnosis
and condition. All examinations were done with dilated pupils.

RESULTS

Table 1 shows the admitting diagnosis of the 82 patients examined. Prematurity and respiratory distress syndrome accounted for 56% of the patients.
The next largest group was composed of children with congenital defects; a
breakdown of this group is in Table 2. The composition of the TPN solution

*Supported in part by a grant from The National Foundation — March of Dimes, the
Bidwill Research Fund and the Cardinal Glennon Research Fund.

Birth Defects: Original Article Series, Volume XII, Number 5, pages 39—43
© 1976 The National Foundation

at Cardinal Glennon Memorial Hospital is as follows: protein hydrolysate (2%), glucose (12.5%), potassium (18.1 meq/1), sodium (14 meq/1), calcium (15.5 meq/1), magnesium (4.8 meq/1), copper (0.0052 meq/1), chloride (8 meq/1), phosphate (22.5 meq/1), zinc (0.0069 meq/1). A multivitamin preparation (1 ml/day) is also added (MVI Concentrate, U. S. Vitamin).

For purposes of analysis, patients were divided into two groups according to whether or not they had received TPN (Table 3). Thirty-one patients received TPN. Of these, 3 developed cortical cataracts (Fig. 1); none had nuclear cataracts. They had received TPN for 3, 5, and 8 weeks. All were admitted as newborns, 1 or 2 days old; all were less than 2 months of age when cataracts were detected. In 51 patients not receiving TPN, 1 patient had an intumescent lens at time of admission; another had sutural cataracts at time of admission and 1 patient developed retrolental fibroplasia.

The patients with cataracts receiving TPN had diagnoses of tracheoesophageal fistula, diaphragmatic hernia and necrotizing enterocolitis. The first 2 had undergone surgery with general anesthesia. Although 2 other patients developed cataracts, neither were cortical. The patient with an intumescent lens had complete cataracts from birth and was admitted with severe sepsis and prematurity. The patient with a sutural cataract had a cleft palate and diarrhea.

TABLE 1.

Prematurity and respiratory distress syndrome	17
Prematurity	6
Respiratory distress syndrome	23
Sepsis	9
Congenital defects	13
Hyperbilirubinemia	7
Convulsive disorder	2
Hypoglycemia	1
Diabetic progeny	2
Necrotizing enterocolitis	2
Total	82

TABLE 2. Congenital Defects

Congenital heart disease	5
Cleft lip/palate	2
Tracheoesophageal fistula	2
Meningocele	1
Gastroschisis	1
Hydrocephalus	1
Rectal stenosis	1
Total	13

TABLE 3. Summary of Patients Examined

Patients on TPN		31
Cortical cataracts	3	
Nuclear	0	
Patients not on TPN		51
Intumescent lens	1	
Sutural cataracts	1	
Retrolental fibroplasia	1	
Total		82

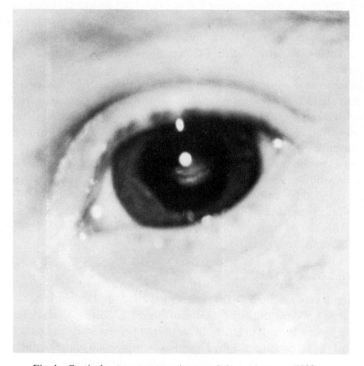

Fig. 1. Cortical cataract as seen in one of the patients on TPN.

DISCUSSION

LeFebvre and Merlen (1) found one congenital cataract in 2153 normal pregnancies, an incidence of 0.04%. In an analysis of 255 cases of congenital cataracts, François (2) found 25% to be hereditary and 10% to be caused by an infectious organism; no etiology was determined in 63%. Spontaneously reversible cataracts are seen in 3% of all premature infants (3, 4). Congenital cataracts are also a feature of a number of multiple congenital defect syndromes. An excellent review of these causes has been published by Kirsch (5).

None of the 3 patients with cataracts receiving TPN in this study had a family history suggesting hereditary origin; none were of the infectious type and none were the spontaneously reversible cataracts, which are seen in premature infants. None of the patients had a multiple congenital defect syndrome.

The numerous complications of TPN have been well reviewed (6–11). Therapeutic problems encountered in our patients during TPN therapy were: alkalosis, acidosis, renal failure, hyponatremia, hyperglycemia, hypocalcemia and hypomagnesemia, hyperammonemia and lactic acidosis. Hyperglycemia, a common complication has been implicated in cataract formation (5). Lactic acidosis has also been implicated in congenital cataracts (12). Though we did not document hyperaminoacidemia, this has been shown in TPN therapy and may play a role (13). What role the other conditions might play must be determined.

Since these were cortical cataracts with no nuclear cataracts and were acquired postnatally, their formation may be related to the TPN therapy. Surgery or anesthesia seems an unlikely causal factor, since many of the patients examined had received surgery and anesthesia and one patient with cataracts had not received anesthesia.

The development of cortical cataracts in patients on TPN merits further intensive evaluation to determine this suggestive association. The current study is being expanded to better evaluate the many variables involved.

In view of the present study, it is advisable that persons caring for patients receiving TPN should be aware of the possible relationship between TPN and cortical cataract formation.

ACKNOWLEDGMENTS

The authors are indebted to Drs. J. E. Lewis and S. F. Bowen, Jr. for allowing us to see some of their patients; to Drs. M. S. Sarnat, V. H. Peden and P. A. Byrne for their help; to Dr. A. E. McElfresh for reviewing and editing the manuscript; to Dr. Anwar Shah for preparing the slides and photographs; and to Miss Doris Weiland and Mrs. Catherine Camp for their help in preparing the manuscript.

REFERENCES

1. LeFebvre, G. and Merlen, J.: La place de la rubéole et des autres facteurs infectieux ou toxiques survenus encours de gestation dans la genèse des malformations et dystrophies congénitales. Ann. Paediat. 171:266, 1948.
2. François, J.: The etiology of congenital cataract. Vie Can. Fr. 43:1327, 1962.
3. Levy, N. S.: Etiology and management of infantile cataracts. Ophthalmol. Digest 35:41, 1973.

4. Yanoff, M., Fine, B. S. and Schaffer, D. B.: Histopathology of transient neonatal lens vacuoles. Am. J. Ophthalmol. 76:363, 1973.

5. Kirsch, R. E.: The lens. Arch. Ophthalmol. 93:284, 1975.

6. Groff, D. B.: Complications of intravenous hyperalimentation in newborns and infants. J. Pediatr. Surg. 4:460, 1969.

7. Heird, W. C., Driscoll, J. M., Schullinger, J. N. et al: Intravenous alimentation in pediatric patients. J. Pediatr. 80:351, 1972.

8. Kaplan, M. S., Mares, A., Quintanca, P. et al: High caloric glucosenitrogen infusions. Arch. Surg. 99:567, 1969.

9. Johnson, J. D., Albritton, W. L. and Sunshine, P.: Hyperammonemia accompanying parenteral nutrition in newborn infants. J. Pediatr. 81:154, 1972.

10. Touloukian, R. J. and Downing, S. E.: Cholestatsis associated with long-term parenteral hyperalimentation. Arch. Surg. 106:58, 1973.

11. Touloukian, R. J.: Isosmolar coma during parenteral alimentation with protein hydrolysate in excess of 4 gm/kg/day. J. Pediatr. 86:270, 1975.

12. Sengers, R. C. A., ter Haar, B. G. A., Trijbels, J. M. F. et al: Congenital cataract and mitochondrial myopathy of skeletal and heart muscle associated with lactic acidosis after exercise. J. Pediatr. 86:873, 1975.

13. Abitbol, C. L., Feldman, D. B., Ahmann, P. and Rudman, D.: Plasma amino acid patterns during supplemental intravenous nutrition of low-birth-weight infants. J. Pediatr. 86:766–772, 1975.

Sacral Agenesis and Associated Anomalies

Jeannette N. Israel, MD , Donald W. Day, MD , Asna Hirschmann, MD
and George F. Smith, MD

Sacral agenesis with its associated anomalies is an interesting birth defect. The absence of the lower segments of the spinal column may not be as uncommon a disorder as once was considered. It was first reported by Hohl (1) in 1852, Wertheim (2) in 1857, and since that time, over 265 cases (3–14) have been reported. Excellent summaries have been published (15) and a detailed survey of 700 children's roentgenograms by Shands and Bundens in Wilmington, Delaware showed an incidence of 0.43% (3, 12).

The agenesis may involve only the terminal elements of the spine and result in few or no symptoms. On the other hand, extensive deletions of the terminal spine may be associated with severe physical dysfunction and a shortened survival. In some cases other congenital anomalies may also be present (16).

The early recognition of sacral agenesis may prevent permanent damage to the urinary tract and kidneys. Unfortunately, diagnosis is seldom made in the newborn period when corrective and preventive measures could be taken. There are a number of findings that should alert the clinician to suspect the existence of this spinal column abnormality. An x-ray examination of the lower spine is a simple confirmatory procedure. A patient with sacral agenesis may have muscular atrophy of the lower limbs, flattening of the buttocks, talipes equinovara of a paralytic type, flexion abduction contractures of the hips and knees, dislocation of the hips or narrowing of the pelvis. A meningomyelocele or spina bifida has also been reported with the condition. On palpation there is loss of the external sacrococcygeal prominence. The patient may stand or ambulate with difficulty or sit in a Buddha position. Neurologically the patient may have absent deep tendon reflexes, paralysis and loss of sensation in the lower limbs. Urinary and bowel incontinence is very frequently present. Urinary

Birth Defects: Original Article Series, Volume XII, Number 5, pages 45–51
© 1976 The National Foundation

tract symptoms may include recurrent urinary tract infection, renal insufficiency and severe incontinence. The IVP may reveal reflux; hydronephrosis; diverticula; stricture; neurogenic bladder; double ureter; or horseshoe, fused, ectopic or absent kidney. As for the GI tract, the patient may have imperforate anus, fecal incontinence, constipation, anal stricture or intestinal malrotation. The following findings have also occurred in patients with sacral agenesis: cleft lip and palate, coarctation of the aorta, diaphragmatic hernia, tracheoesophageal fistula and agenesis of the lung (17).

We have seen 3 patients with sacral agenesis in the past year, 2 of whom were newborns. The first newborn had an unusual and severe form of the anomaly.

CASE REPORTS

Case 1

The patient, a black male, was born at 30 weeks' gestation to a 17-year-old gravida 1, para 0 mother (Fig. 1). The pregnancy was normal except for a "cold" in the second month. There was no history of diabetes mellitus, fever or chemical

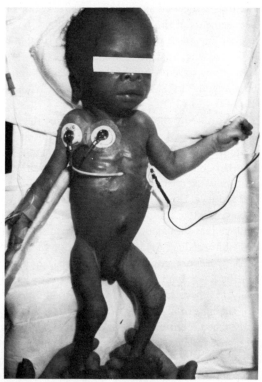

Fig. 1. *Case 1* showing body size and enlarged head.

exposure during the pregnancy. The membranes had ruptured prematurely with the passage of 100 cc of dark red blood. The infant's Apgar score was low and he required resuscitation. The infant weighed 1360 gm and was in moderate respiratory distress. There was a skull defect in the parietal area and the skull bones were widely separated. The ears were low-set and malformed. The heart and abdomen were normal. The muscles of the lower limbs were atrophic and the feet deviated medially. The rectal sphincter tone was poor. Radiographs of the spine showed agenesis of the sacrum below S1 and hypoplasia of the lumbar spine (Fig. 2). At 3 days of age, the patient developed tremors due to hypocalcemia. Petechiae were also noted at that time and the CBC showed pancytopenia with a Hb of 11.3, WBC of 2900 and a platelet count of 40,000. Despite supportive measures the patient expired at 5 days of age.

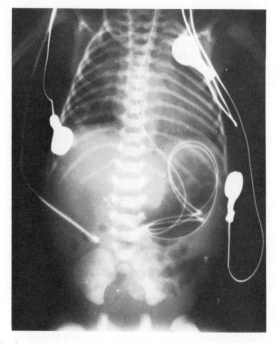

Fig. 2. Radiograph of *Case 1* showing absence of sacral bones.

An evaluation of the patient showed no evidence of a prenatal infection. A bone marrow sample taken shortly before death was unsatisfactory for diagnostic purposes. Placental examination showed a large area of infarction and an area of recent hemorrhage. The autopsy revealed an internal communicating hydrocephalus with multiple foci of intracerebral hemorrhage and there was a vascular malformation of the cerebellum. The lungs showed acute bronchopneumonia probably due to aspiration. There was agenesis of the sacrum and hypoplasia of the spines of the lumbar vertebras. The kidneys were normal as was the rest of the GU system.

This patient presented the unusual findings of sacral agenesis in association with internal communication hydrocephalus and vascular malformation of the cerebellum. He had unusual-appearing ears and a skull defect. The prenatal history was negative for any of the commonly speculated causative factors for this syndrome, eg diabetes, vitamin A deficiency, high fever, maternal exposure to fat solvents or ingestion of lithium salts or aminonicotinamide. The mother has a retarded brother who does not have either sacral agenesis or hydrocephalus.

Case 2

The second patient was a white female newborn. She was found to have bilateral clubfeet and sacral agenesis at birth. The maternal history was negative for diabetes mellitus and chemical exposure. Bilateral corrective casts were applied to the feet in the nursery.

At 3 weeks of age, the patient was readmitted to the hospital with low urine output and fever. On physical exam, the bladder was distended and there was a large 3 X 4 cm mass in the left upper quadrant of the abdomen. Cystoscope revealed a large left hydronephrosis with segmental stricture of the left ureter. A left cutaneous ureterostomy was performed and the postoperative course was uneventful. The patient is presently being followed as an outpatient.

Case 3

The third patient was an 8-year-old white female and a known case of sacral agenesis when she was admitted to our hospital for orthopedic surgery for pes cavus. She had been experiencing incontinence but was felt to be doing well otherwise. Other prior history is unavailable because the child is with foster parents.

The physical exam showed paralytic pes cavus and no gluteus maximus musculature. The patient had an abnormal gait chracterized by bilateral talipes cavovarus, right greater than left, and external femoral rotation. There was atrophy and weakness of the calf muscles. The pes cavovarus was characterized by hindfoot calcaneus, forefoot equinus and plantar fascia contracture.

The radiographs demonstrated 4 typical lumbar vertebras and complete absence of the sacrum (Fig. 3). The 5th lumbar vertebra had an elongated narrow body. The posterior neural arch of L5 was deformed and articulated with the ileum bilaterally. The iliac bones appeared normal but were close together, resulting in considerable narrowing of the transverse diameter of the pelvis. The hip joints were similarly close together but were otherwise normal. The IVP showed both kidneys normal with delayed and poor emptying of contrast material which is consistent with neurogenic bladder. There was mild right hydroureter which may be due to nonspecific inflammation. Corrective surgical procedures were performed for the congenital deformities of the feet.

DISCUSSION

An evaluation of the data on sacral agenesis suggests that the condition, at times, can be considered a separate disease entity, but it can also occur as a component of a more complex disease process as demonstrated in *Case 1*. If sacral agenesis is not a single entity, then we would expect that different etio-

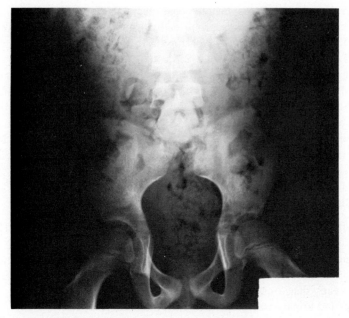

Fig. 3. Radiograph of *Case 3* showing pelvis and lumbar regions and demonstrating absence of sacral and coccyx bones.

logic mechanisms would be producing similar sacral anomalies and this is exactly what appears to be happening in this condition. For example, most cases of sacral agenesis have been sporadic with a few instances of familial occurrence. Pouzet (18) reported a case of father and son with absence of the sacrum. Kenefick (8) reported a family with 6 females and 3 males with sacral agenesis. Cohn and Bay-Nielsen (6) reported one family in which there were 6 females with sacral agenesis and anterior sacral meningocele, suggesting X-linked dominant inheritance. Dunn and co-workers (19, 20) described a new gene in mice producing deletion defects of the spinal column and GU abnormalities. Danforth reported genetic rumplessness in fowl (5, 21). There does not appear to be an exact counterpart in man for the latter abnormality.

An important observation is the relation of maternal diabetes to sacral agenesis. Passarge and Lenz (7) estimated that 16% of mothers of infants with sacral agenesis had a history of diabetes. This correlation was noted by Pedersen et al (22) and has also been confirmed by other investigators (20, 23–26).

There have been many animal studies with exogenous substances to produce the defect of sacral agenesis. The following conditions have produced sacral agenesis in animal embryos: high temperatures — Danforth (5, 21); fat solvents — Kucera (5, 9); hypoglycemia produced by insulin injection — Duraiswami (5, 27); 6-aminonicotinamide — Nogami and Ingalls (5, 28); lithium — Lehmann (5, 29); and vitamin A deficiency — Zilva et al (5, 30).

The theory of inductive interaction suggests a mechanism by which sacral agenesis can occur. The theory proposes that the formation of a particular specialized tissue depends on the transmission of stimuli from adjacent tissue. Sacral agenesis may therefore, be due to a failure of interaction between notocord and neural ectoderm (5). In the human embryo, differentiation of lumbar vertebras, sacrum and coccyx occurs between the fourth and seventh weeks of gestation and the defect would be expected to occur during this time (31).

CONCLUSION

Sacral agenesis can no longer be considered an uncommon disorder. The condition occurs often enough that the clinician should be alert to any signs that may suggest its presence in a newborn. This disease is frequently associated with other anomalies, especially those of musculoskeletal, GU and GI systems. Diagnosis — early diagnosis — is the key to prevention of permanent damage to the kidneys and urinary tract in patients with the disease.

ACKNOWLEDGMENT

Appreciation is acknowledged for the assistance of Dr. R. Clausen who performed the pathologic examination on *Case 1*.

REFERENCES

1. Hohl, A. F.: "Zur Pathologie des Beckens." Leipzig: W. Englemann, p. 61.
2. Wertheim, C. C.: Vollständiger Mangel des Kreuz-und Steissbins bei einem Neugeborenen. Monatschr. Geburtsk. Frauenkr. 9:127, 1857.
3. Thompson, I. M., Kirk, R. M. and Dale, M.: Sacral agenesis. Pediatrics 54:236, 1974.
4. Lourie, H.: Sacral agenesis. J. Neurosurg. 38:93, 1973.
5. Marsh, H. O. and Tejano, N. A.: Four cases of lumbosacral and sacral agenesis. Clinical Orthop. 92:214, 1973.
6. Cohn, J. and Bay-Nielsen, E.: Hereditary defects of the sacrum and coccyx with anterior sacral meningocele. Acta Paediatr. Scand. 58:268, 1969.
7. Passarge, E. and Lenz, W.: Syndrome of caudal regression in infants of diabetic mothers: Observations of further cases. Pediatrics 37:672, 1966.
8. Kenefick, J.: Hereditary sacral agenesis associated with presacral tumours. Br. J. Surg. 60:272, 1973.
9. Kucera, J.: Exposure to fat solvents: A possible cause of sacral agenesis in man. J. Pediatr. 72:857, 1963.
10. Koontz, W. and Prout, G.: Agenesis of the sacrum and neurogenic bladder. JAMA 203: 481, 1968.
11. Alexander, E. and Nashold, B.: Agenesis of the sacro-coccygeal region, case report and technical notes. J. Neurosurg. 13:507, 1956.

12. Shands, A. R. and Burdens, W. D.: Congenital deformities of the spine and analysis of the roentgenograms of 700 children. Bull. Hosp. Joint Dis. 17:110, 1956.

13. Frantz, C. H. and Aitkens, G. T.: Complete absence of lumbar spine and sacrum. J. Bone Joint Surg. 49A:1531, 1967.

14. Rusnak, S. L. and Driscoll, S. G.: Congenital spine anomalies in infants of diabetic mothers. Pediatrics 35:989, 1965.

15. Freeman, B.: Congenital absence of the sacrum and coccyx: Report of a case and review of the literature. Br. J. Surg. 37:299, 1950.

16. Bell, J. F., Kuhlmann, R. F. and Molloy, M. K.: Congenital defects of shoulder girdle, sternum, spine and pelvis. Pediatr. Clin. North Am. 14:413, 1967.

17. Smith, E. D.: Congenital sacral defects. In Stephens, F. D. (ed.): "Congenital Sacral Defects in Congenital Malformations of Rectum, Anus, and Genito-urinary Tracts," London: E. and S. Livingstone, Ltd., 1963, p. 82.

18. Pouzet, F.: Les anomalies de development du sacrum. Lyon Chirurg. 35:371, 1938.

19. Dunn, L. C., Gluecksohn-Schoenheimer, S. and Bryson, V.: A new mutation in the mouse. J. Hered. 31:343, 1940.

20. Blumel, J., Evans, E. A. and Eggers, G. W.: Partial and complete agenesis or malformation of the sacrum with associated anomalies, etiology and clinical study with special reference to heredity. A preliminary report. J. Bone Joint Surg. 41A: 497, 1959.

21. Danforth, C. H.: Artificial hereditary suppression of the sacral vertebrae in the fowl. Proc. Soc. Exp. Biol. Med. 30:143, 1940.

22. Pedersen, L. M., Tygstrup, I. and Pedersen, J.: Congenital malformations in newborn infants of diabetic women — correlation with maternal vascular complications. Lancet 1:1124, 1964.

23. Banta, J. V. and Nichols, O.: Sacral agenesis. J. Bone Joint Surg. 51A:693, 1969.

24. Stern, L., Ramos, A. and Light, L.: Sacral agenesis in infants of diabetic mothers. Lancet 1:1393, 1965.

25. Passarge, E.: Congenital malformations and maternal diabetes. Lancet 1:324, 1965.

26. Russel, H. E. and Aitkens, G. T.: Congenital absence of the sacrum and lumbar vertebrae with prosthetic management. A survey of the literature and presentation of five cases. J. Bone Joint Surg. 45A:501, 1963.

27. Duraiswami, P. K.: Insulin induced skeletal abnormalities in developing chickens. Br. Med. J. 2:384, 1950.

28. Nogami, H. and Ingalls, T.: Pathogenesis of spinal malformations induced in the embryos of mice. J. Bone Joint Surg. 49A:1551, 1967.

29. Lehmann, F. E.: Die Entwicklung von Ruckenmark, Spenal-ganglien und wirbelanlagen in chordae. Rev. Suisse Zool. 42:405, 1935.

30. Zilva, S. S., Golding, J., Drummond, J. C. and Coward, K. K.: Relation of the fat solvent factors to rickets and growth in pigs. Biochem. J. 15:427, 1921.

31. Arey, L. B.: "Developmental Anatomy: A Textbook and Laboratory Manual of Embryology." Philadelphia: W. B. Saunders, 1957, p. 407.

Monozygotic Twinning and the Duhamel Anomalad (Imperforate Anus to Sirenomelia): A Nonrandom Association Between Two Aberrations in Morphogenesis*

David W. Smith, MD, Catherine Bartlett and Lyle M. Harrah

The concept of the Duhamel anomalad, from imperforate anus to sireno-melia, as representing a single, early, localized anomaly of varying severity with sirenomelia (mermaid-limb) being the most severe expression, has been set forth by Smith et al (1). Less severe and more common instances include many patients with the VATER association (2, 3) for which the emphasis is on vertebral ano-malies, anal atresia, esophageal atresia with tracheoesophageal fistula (TEF), radial limb anomaly and renal defect; the caudal regression syndrome (4, 5) which emphasizes the lower vertebral and pelvic anomalies; the Rokitansky syndrome (6, 7) for which the main feature is incomplete morphogenesis of the uterus and upper vagina; and other nonrandom associations of defects from the Duhamel anomalad. A striking association has been recognized between the severe degree of the Duhamel anomalad (sirenomelia) and monozygotic twin-ning (8). The purpose of this report is to explore the relationship between vary-ing degrees of the Duhamel anomalad and monozygotic twinning in an effort to establish its relevance to the nature and etiology of both monozygotic twinning and the Duhamel anomalad.

METHODS

The constituent malformations which comprise the Duhamel anomalad may variably include partial-to-complete fusion of the lower limbs (sirenomelia), vertebral and/or pelvic anomalies, imperforate anus, renal agenesis or dysplasia,

*Supported by Maternal and Child Health Services, Health Services and Mental Administra-tion, Department of HEW, Project 913; NIH grant HD 05961; USPHS grant GM 15253; and The National Foundation – March of Dimes.

Birth Defects: Original Article Series, Volume XII, Number 5, pages 53–63

incomplete uterine and/or vaginal morphogenesis, defects of external genitalia, single umbilical artery and, occasionally, esophageal atresia, radial limb hypoplasia and defects of sidedness.

The literature was reviewed in order to ascertain both the frequency of monozygous twinning among patients with each of these individual defects and the frequency of these various components of the Duhamel anomalad in monozygotic vs dizygotic twins. A search was also made for instances of malformations among the co-twins of cases of sirenomelia.

In addition, one of the authors (C.B.) evaluated the hospital records of 506 patients at the Children's Orthopedic and University Hospitals who had one or more of the constituent defects of the Duhamel anomalad. Six percent of the cases were doubly ascertained. At least 32% had multiple malformations. The presence or absence of twinning was sought from the hospital record. When a patient was found to be a twin, the presumed zygosity was determined from data in the hospital record, obstetric and newborn records, from private physicians and from the families. Zygosity judgments were based on a combination of data on placenta, membranes and phenotype.

A search was also made of our own records for instances in which a monozygotic twin was affected with the Duhamel anomalad.

RESULTS

Frequency of Presumed Monozygotic Twinning in Patients with the Duhamel Anomalad

Table 1 summarizes the data which shows an excess of monozygotic twins among patients with sirenomelia as well as those ascertained by virtue of a single feature of the Duhamel anomalad. Davies et al (8) noted the striking excess of presumed monozygotic twinning in cases of sirenomelia. It should be noted that only 2 (9%) of the 23 monozygotic twin sirenomelia cases were concordant for this anomalad. Our own study revealed 9 presumed monozygotic twins with unaffected co-twins among 506 patients ascertained on the basis of a single defect of the Duhamel anomalad. This is 2.1 times the expected frequency for monozygous twins, whereas the number of presumed dizygotic twins was the expected number of 4. Of the 9 affected monozygous twins found in this study, 3 had a second malformation; in each case it was of the Duhamel anomalad type. One patient had renal agenesis and imperforate anus; the second had renal agenesis and esophageal atresia; and the third had unilateral renal agenesis plus a bicornuate uterus. David and O'Callaghan (9) found a 6% frequency of twinning in patients ascertained as having esophageal atresia with TEF fistula, and estimated that about half of the cases were monozygotic. An increased frequency of monozygotic twinning has also been found in patients ascertained as having a single umbilical artery (10).

TABLE 1. Frequency of Presumed Monozygous Twinning in Patients Ascertained as Having One or More Componets of the Duhamel Anomalad

Ascertainment for	Source	Number of Patients	Presumed Mono- zygous Twinning	Relative Increase*
Sirenomelia	Davies et al (8)	327	7.0%	8×
Imperforate anus	Present study	175	1.7%	2×
Renal agenesis or dysplasia	Present study	161	2.4%	2.7×
Bicornuate uterus or septate vagina	Present study	69	1.4%	1.6×
Esophageal atresia with TEF	Present study	132	1.5%	1.7×
Esophageal atresia with TEF	David et al (9)	324	about 3.0%	3.4×
Single umbilical artery	Hyrtl (10)	15	20%	23×

*Expected incidence of monozygous twin births in the general population is 0.44% and the frequency of monozygotic twins is thereby 0.88% in the population.

Frequency of the Duhamel Anomalad in Monozygotic Twins

Two of the 5 cases with sirenomelia included in the WHO survey of the off-spring of 421,781 pregnancies (11) were one of presumed monozygotic twins. In that study sirenomelia was about 150 times more frequent in one of the monozygotic twins than in singletons.

The incidence of individual features of the Duhamel anomalad in patients who did not have sirenomelia is also increased in monozygotic twins, as indicated by the data of the NIH Collaborative Perinatal Study on the offspring of 56,000 pregnancies (12). The pertinent data is summarized in Table 2. From autopsy studies Hyrtl (10) noted a single umbilical artery in 7% of monozygotic twins, a sevenfold increase over the expectancy in singletons.

In about 1% of monozygotic twins, one of them is only partially represented by varying degress of incomplete-to-amorphous development, often termed acardiac monster (13—16). Of interest is the fact that many of these cases have defects of the Duhamel anomalad type, including sirenomelia (15). A detailed review of 23 cases in the literature which had complete pathological examination disclosed 4 with sirenomelia, 7 with 2 or more of the Duhamel anomalad defects and 3 with one of this pattern of defects.* Thus at least 14 of these 23 acardiac-to-amorphous monozygotic twins had varying degrees of the Duhamel anomalad.

*References available from the authors on request.

TABLE 2. Frequency of Duhamel Anomalad Type of Defects in One of the Monozygotic Twins vs Singletons in the NIH Collaborative Perinatal Study* (12)

Features	Singletons Rate per 1000	373 Monozygotic Twins Rate per 1000	Relative Increase over Singleton Rate
Renal agenesis or cystic kidney	0.82	5.2	6.3×
Imperforate anus	0.46	2.6	5.6×
Esophageal atresia with TEF	0.18	2.6	14.4×
Total for 3 defects	1.46	10.4	7.1×

*There was no instance of these defects among 617 dizygotic twin pairs.

Another subgroup of monozygotic twins is the conjoined twins (13, 14, 17–19), who represent about 0.5% of monozygotic twinning. Such twins may be joined at a number of different sites as a result of overlapping fields of morphogenesis, most commonly at the thorax, eg the thoracopagus conjoined twin. The Duhamel anomalad is also a frequent finding in these patients. Detailed review of 12 instances of conjoined twins disclosed one pair discordant for sirenomelia (20), one concordant for imperforate anus and bicornuate uterus plus septate vagina (21), and one discordant for unilateral absence of the radius and thumb (22). Of interest as a link between the amorphous twin and conjoined twins was an instance of thoracopagus, one of which was incompletely developed with no heart.

Defects in Co-Twins of a Monozygotic Twin with the Duhamel Anomalad

We found the following features in 5 cases of partially affected presumed monozygotic co-twins of patients with sirenomelia: 1) unilateral renal agenesis (23); 2) unilateral renal agenesis and imperforate anus (24); 3) renal anomaly, imperforate anus and cryptorchidism (24); 4) cystic fused kidney and single umbilical artery (25); 5) agenesis of external genitalia and asymmetry of limbs (11). All cases have a less severe degree of the Duhamel anomalad than the sirenomelia twin.

Table 3 summarizes the findings in six nonconcordant presumed monozygotic twins with 2 or more defects of the Duhamel anomalad.

DISCUSSION

The findings of this study further establish a relationship between monozygotic twinning and the Duhamel anomalad. Monozygotic twinning is an aberration in morphogenesis and the frequency of associated defects is twice as high as in either dizygotic twins or in singletons (11, 12). Unfortunately, because

TABLE 3. Presumed Monozygotic Twins With More Than One of the Defects of the Duhamel Anomalad

Reference	Imperforate Anus	Renal Agenesis or Dysplasia	Bicornuate Uterus or Septate Vagina	Cardiac Defect	Other
Personal case*	+		+		Diastasis of pubis Persistent urachus
Personal case	+			+	Common cloaca Sacral spina bifida Right equinovarus Aplasia of left tibia
26	+	+		+	Esophageal atresia Hemivertebrae Clubfoot Hydrocephalus with Arnold-Chiari
27	+	+		+	Hypoplasia of pelvis and fibulas Absent spleen Four-lobed lungs
28	+		+		
29	+	+	+		

*Courtesy of Col. James L. Stewart, Jr., Letterman General Hospital, San Francisco, California.

of the tendency to think of monozygotic twins as identical, there is a natural reluctance to diagnose monozygosity when only one of the twins has a malformation. Actually, the majority of monozygous twins with a malformation problem of the single primary anomaly type are discordant (30). Since monozygotic twins are twice as likely to have a malformation as dizygotic twins, a twin pair with a malformation in one of them is actually more likely to be monozygotic than dizygotic.

At least part of the excess frequency of malformation in monozygous twins is due to the increased likelihood of having the Duhamel anomalad, each feature of which is more than 3 times as common in monozygotic twins as in singletons or dizygotic twins (Table 2). Also, the Duhamel anomalad may be responsible for some of the increased mortality rate in monozygotic twins, a mortality which is 27% greater than in dizygotic twins (13).

Although the precise causes for monozygotic twinning and for the Duhamel anomalad are unknown, their association suggests an etiologic relationship. Prior to setting forth a hypothesis as to the reason for this association, some relevant background information on the developmental pathogenesis of both abnormalities will be presented.

There are probably many causes for monozygotic twinning (Fig. 1), since it can occur at any stage until after the primitive streak (11–13). In about one third of the instances the inception of the defect is during the first 4–5 days, prior to blastocyst formation (13). This can give rise to 2 blastocysts and thereby 2 chorions (dichorionic) and 2 amnions (diamniotic). The wall of the blastocyst gives rise to the chorion. Such monozygous twins can even have separate placentas. The most common type of monozygotic twinning appears to be the result of 2 embryonic centers (germinal centers) being organized in one blastocyst. This must occur between the time the blastocyst is formed at 4–5 days and the time the amnion has developed from the embryonic center at 8–9 days (13). Such an example was beautifully demonstrated by Assheton (31) in a 7-day sheep blastocyst which had 2 embryonic centers in its wall rather than one. If one germinal center was rudimentary, such as reported by Streeter in a 17-day human (32), the result might be an incomplete-to-amorphous twin. If the germinal centers were close together with overlapping fields of morphogenesis, the result might be conjoined twins in a single amnion. Should the embryonic centers be separate, the result would be monochorionic-diamniotic twins, the most common monozygotic type (33). The final type, with one chorion and one amnion, presumably has its onset in the embryonic disk after the development of the amnion and before full development of the primitive streak, between 8 and 16 days. This could arise by a division of the embryonic disk or by the organization of more than one primitive streak within the same germinal disk. If these primitive streaks were partially joined or too close together, the result could be conjoined twins.

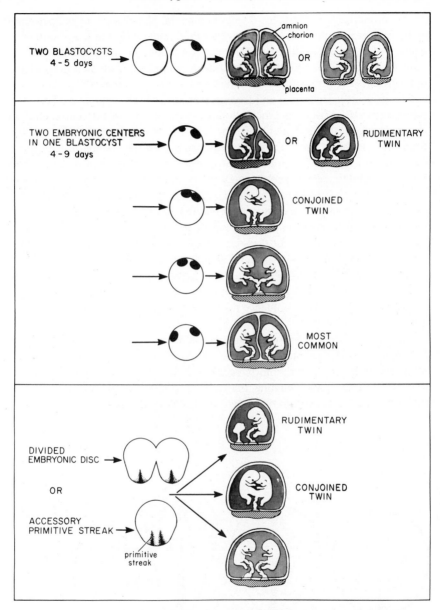

Fig. 1. Developmental pathogeneses for monozygotic twinning. The heavy black line represents the chorion, derived from the wall of the blastocyst, and the light lines represent the amnion, derived from the embryonic center. (Adapted from Bulmer [13] and Benirschke and Kim [14].)

One attractive hypothesis for the developmental pathogenesis of the pattern of malformation in Duhamel anomalad is that there was a defect in the primitive streak with a secondary deficit in the migration of mesoderm, especially the caudal mesoderm. The primitive streak, the development of which is illustrated in Figure 2, is the first structure to provide an axial orientation to the early embryo (34–38). Through it and from it, mesodermal cells migrate to their presumptive designations in anticipation of organogenesis. At the central end of the primitive streak the Hensen node forms and from it cells migrate forward to form the notochord, which exerts an inductive role on the development of the overlying neural tube. The primitive streak mesoderm sweeps forward to form the somites, the intermediate (nephrogenic) mesoderm and at least portions of the lateral plate mesoderm. Having played its critical role as the forerunner of organogenesis from 15–16 days until about 18–19 days, the primitive streak recedes. Obviously, errors in its development and function could adversely affect morphogenesis in a number of tissues. A severe degree of defect could result in a gross deficiency of caudal mesoderm with secondary fusion of the lower limb buds, ie sirenomelia. Wolff (39) has produced such a malformation complex by irradiation of the chick primitive streak and caudal mesoderm. Lesser degrees of such an early anomaly could give rise to lesser degrees of the Duhamel anomalad.

The hypothesis for the association between monozygotic twinning and the Duhamel anomalad is depicted in Figure 3 and is as follows: Whatever the cause of the aberration in the early morphogenesis of the embryonic center(s) which resulted in monozygotic twinning, the same cause may give rise to a problem in the early organization within the embryonic disk(s) and adversely affect primitive

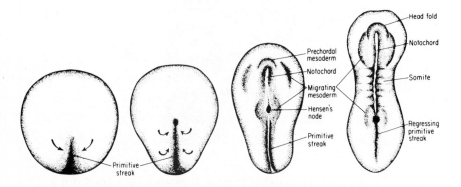

Fig. 2. Representation of primitive streak development and regression from about 15 days to about 18–19 days. (Adapted from photos of early human development by Nishimura [34] plus photos of early chick development from Hamburger and Hamilton [35].)

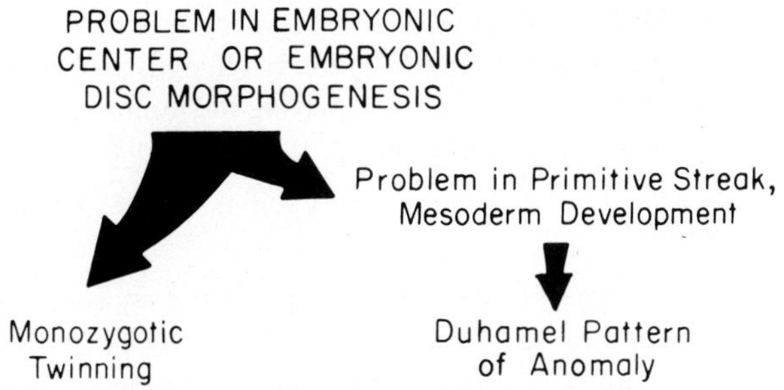

PROBLEM IN EMBRYONIC
CENTER OR EMBRYONIC
DISC MORPHOGENESIS

Problem in Primitive Streak,
Mesoderm Development

Monozygotic
Twinning

Duhamel Pattern
of Anomaly

Fig. 3. Possible developmental pathogenesis for the association between monozygotic twinning and the Duhamel anomalad.

streak formation and function, thereby resulting in an increased likelihood of the Duhamel anomalad. This would be especially true for the embryonic centers which are rudimentary and give rise to incompletely developed acardiac and/or amorphous co-twins, and for embryonic centers which are partially fused or those in which more than one primitive streak developed, giving rise to conjoined twins. Both of these latter types have a relatively high frequency of the Duhamel anomalad.

One other anomaly which has a high frequency in monozygotic twins (22%) and an even higher incidence in conjoined twins (73%) is situs inversus viscerum (26). Defects of sidedness, including bilateral right sidedness (asplenia syndrome, Ivemark syndrome) (40) and bilateral left sidedness (polysplenia syndrome) (40) appear to be more common in patients with the Duhamel anomalad (27, 41, 42). The period when the primitive streak develops is also the period when the axis of symmetry is established, including positioning and polarity (37, 38). Hence, extending the previous hypothesis, a problem which upsets the organization of the embryonic center or disk may give rise to monozygotic twinning; it may also give rise to a problem in primitive streak mesoderm development, resulting in the Duhamel anomalad, and may also affect the axis of symmetry and result in asymmetries.

The findings in monozygotic twins provide additional strength to the concept of the Duhamel anomalad as a single early defect of varying severity. This concept is supported by the 5 partially affected presumed monozygotic co-twins of sirenomelia cases. In all 5 cases the defects in the co-twin are of the Duhamel anomalad type and spectrum. At least several of them might have been diagnosed as having the VATER association (2). The data in Table 3 also provide support for this concept.

The following conclusions seem justified: 1) One of the reasons for an increased frequency of associated malformations in monozygotic twins is the increased frequency of the Duhamel anomalad in monozygotic twins. 2) Clinicians should be more alert than usual to the detection of the Duhamel anomalad in monozygotic twins. 3) The basic cause of the problem in embryonic center or disk development which results in monozygotic twinning is probably also the cause for the Duhamel anomalad in patients having the nonrandom association of both of these problems in morphogenesis.

ACKNOWLEDGMENTS

We wish to acknowledge Mrs. Mary Ann Sedgwick Harvey for editorial assistance, and Mrs. Margaret Lankford of the Children's Orthopedic Hospital record room and Ms. Anne Gregor of the University Hospitals record room for their participation in this study.

REFERENCES

1. Smith, D. W., Barlett, C., Hanson, J. W. and Harrah, L. M.: The Duhamel anomalad, from imperforate anus to sirenomelia; including the VATER association, caudal regression syndrome, and Rokitansky syndrome. (To be published.)
2. Quan, L. and Smith, D. W.: The VATER association. J. Pediatr. 82:104, 1973.
3. Temtamy, S. A. and Miller, J. D.: Extending the scope of the VATER association: Definition of the VATER syndrome. J. Pediatr. 85:345, 1974.
4. Duhamel, B.: From the mermaid to anal imperforation: The syndrome of caudal regression. Arch. Dis. Child. 36:152, 1961.
5. Kučera, J., Lenz, W. and Maier, W.: Malformations of the lower limbs and the caudal part of the spinal column in children of diabetic mothers. Ger. Med. Mon. 10:393, 1965.
6. Bryan, A. L., Nigro, J. A. and Counseller, V. S.: One hundred cases of congenital absence of the vagina. Surg. Gynecol. Obstet. 88:79, 1949.
7. Van Campenhout, J. and Leduc, B.: Unusual features in the Rokitansky-Küster-Hauser syndrome. Lancet 2:928, 1971.
8. Davies, T., Chazen, E. and Nance, W. B.: Symmelia in one of monozygotic twins. Teratology 4:367, 1970.
9. David, T. J. and O'Callaghan, S. E.: Twinning and oesophageal atresia. Arch. Dis. Child. 49:660, 1974.
10. Hyrtl, J.: "Die Blutgefässe der menschlichen Nachgeburt in normalen und abnormalen Verhältnissen." Vienna:Braumüller, 1870.
11. Stevenson, A. C., Johnson, H. A., Stewart, M. I. P. and Golding, D. R.: Congenital malformations; a report of a study of series of consecutive births in 24 centres. WHO Bull. (Suppl. 1)34:127, 1966.
12. Myrianthopoulos, N.: Personal communication regarding data from the NIH Collaborative Perinatal Study.
13. Bulmer, M. G.: "The Biology of Twinning in Man," Oxford: Clarendon Press, 1970.
14. Benirschke, K. and Kim, C. K.: Multiple pregnancy. N. Engl. J. Med. 288:1276, 1973.
15. Kappelman, M. D.: Acardius amorphus. Am. J. Obstet. Gynecol. 47:412, 1944.
16. Keith, A.: The anatomy and nature of two acardiac acephalic foetuses. Trans. Obst. Soc. London 42, 1900.
17. Dorland, W. A. N.: Some rare foetal teratisms, with illustrative cases — sympodia, craniopagus and acephalus. Surg. Gynecol. Obstet. 20:342, 1915.
18. Willis, R. A.: "The Borderland of Embryology and Pathology," 2nd Ed. Washington: Butterworth, 1962.

19. Igarashi, M., Singer, D. B., Alford, B. R. and Cook, T. A.: Middle and inner ear anomalies in a conjoined twin. Laryngoscope 84:1188, 1974.
20. Krone, H. A. and Mahdavyan, M.: Beobachtung einer seltenen Sirenen-Doppelmissildung. Beitrag zur kausalen Genese menschlicher Entwicklungsstorungen. Geburtshilfe Frauenheilkd. 19:435, 1959.
21. Borden, S., IV, Rider, R. G., Pollard, J. J. and Hendren, W. H.: Radiology of conjoined twins. Intrauterine diagnosis and postnatal evaluation. Am. J. Roentgenol. Radium Ther. Nucl. Med. 120:424, 1974.
22. Rogala, E. J., Wynne-Davies, R., Littlejohn, A. and Gormley, J.: Congenital limb anomalies: Frequency and aetiological factors. J. Med. Genet. 11:221, 1974.
23. Källen, B. and Winberg, J.: Caudal mesoderm pattern of anomalies: From renal agenesis to sirenomelia. Teratology 9:99, 1974.
24. Bauereisen, T.: Zwilling ist eine typische sirenenbildung Erlangen. Munch. Med. Wochenschr. 52:721, 1905.
25. Faierman, E.: The significance of one umbilical artery. Arch. Dis. Child. 35:285, 1960.
26. Kučera, J. and Lenz, W.: Caudale Regression mit Oesophagusatresie und Nierenagenesis – ein Syndrom. Z. Kinderheilkd. 98:326, 1967.
27. Kohler, H. G.: An unusual case of sirenomelia. Teratology 6:295, 1972.
28. Halbert, D. R. and Christakos, A. C.: Discordance of sexual anomalies in monozygotic twins: Report of a case. Obstet. Gynecol. 36:388, 1970.
29. Guldberg, E.: Verschieden Geshlechtige eigene Zwillinge. Acta Pathol. Microbiol. Scand. Suppl. 37:197, 1938.
30. Smith, D. W.: "Recognizable Patterns of Human Malformation." Philadelphia: W. B. Saunders Company, 1970.
31. Assheton, R.: An account of a blastodermic vesicle of the sheep of the seventh day, with twin germinal areas. J. Anat. Phys. 32:362, 1898.
32. Streeter, G. L.: Formation of single-ovum twins. Bull. Johns Hopkins Hosp. 30:235, 1919.
33. Edwards, J. H.: The value of twins in genetic studies. Proc. R. Soc. Med. 61:227, 1968.
34. Nishimura, H., Tanimura, T., Semba, R. and Uwabe, C.: Normal development of early human embryos: Observation of 90 specimens at Carnegie Stages 7 to 13. Teratology 10:1, 1974.
35. Hamburger, V. and Hamilton, H. L.: A series of normal stages in the development of the chick embryo. J. Morphol. 88:49, 1951.
36. Hunt, T. E.: Potencies of transverse levels of the chick blastoderm in the definitive streak stage. Anat. Rec. 55:41, 1932.
37. Romanoff, A. L.: "The Avian Embryo. Structural and Functional Development." New York: The MacMillan Company, 1960.
38. Langman, J.: "Medical Embryology," 2nd Ed. Baltimore: The Williams and Wilkins Company, 1969.
39. Wolff, E.: Les bases de tératogénèse expérimentale des vertébrés amniotes d'après les résultats de méthodes directes. Arch. Anat. Histol. Embryol. (Strasb.) 22:1, 1936.
40. Van Mierop, L., Gessner, I. H. and Schiebler, G. L.: Asplenia and polysplenia syndromes. In Bergsma, D. (ed.): Part XV. "The Cardiovascular System," Birth Defects: Orig. Art. Ser., vol. VIII, no. 5. Baltimore: The Williams and Wilkins Co. for The National Foundation – March of Dimes, 1972, pp. 36–44.
41. Chandra, R. S.: Biliary atresia and other structural anomalies in the congenital polysplenia syndrome. J. Pediatr. 85:649, 1974.
42. Rusnak, S. L. and Driscoll, S. G.: Congenital spinal anomalies in infants of diabetic mothers. Pediatrics 35:989, 1965.

Aberrant Tissue Bands and Craniofacial Defects*

Diane L. Broome, MD, Allan J. Ebbin, MD, August L. Jung, MD,
L. Richard Feinauer, MD and Michael Madsen

We wish to describe 8 infants with a similar pattern of malformations of craniofacial deformities and aberrant tissue bands of the limbs first reported by Jones and co-workers (1). We are reporting these cases to delineate further this syndrome for improved diagnosis, to describe the variation in clinical prognosis and to demonstrate the effects of reconstructive surgery. The family data and various malformations are summarized in Tables 1 and 2.

CASE REPORTS

Case 1

A 3317-gm male was born at term to a 20-year-old Mexican-American primigravida by C section indicated because of frank breech presentation. The mother had a urinary tract infection in the first month of pregnancy. There was no family history of congenital defects or excess fetal deaths. To date the mother has not had another pregnancy.

Severe birth anomalies included an occipital encephalocele; absence of the anterior portion of the skull; bilateral cleft lip and palate; left anophthalmia; distorting band-like tissue on the face extending from the left upper cheek and the base of the nose on the right to the forehead (Fig. 1); and amputations of the right 3rd, 4th, and 5th fingers at the proximal phalanges, and the left 2nd, 3rd, and 4th fingers at the distal phalanges. The placenta was not examined.

*Supported in part by grant no. 286 from Maternal and Child Health Service, USPHS (Miriam G. Wilson, M.D., Chief Investigator); a contract from the National Institute of Environmental Health Sciences through the American Academy of Pediatrics; and the Los Angeles County Chapter of The National Foundation — March of Dimes (Allan J. Ebbin, M.D., Chief Investigator).

Birth Defects: Original Article Series, Volume XII, Number 5, pages 65–79
© 1976 The National Foundation

TABLE 1. Family Data

Case	Weight (gm)	Gestational age	First live-born	Delivery	Pregnancy history	Maternal age	Paternal age	Sex	Race	Outcome
1	3317	Term	+	C/S	Urinary tract infection	20	22	M	MA	Died at 25½ mo
2	2960	Term (twin)	–	V		31	38	F	MA	Surviving at 2½ yr
3	2155	Term	twins	V	chest pain	17	19	M	B	Died at 24 hr
4	2280	Term	–	V	prior drug use and influenza	21	20	F	W	Died at 24 hr
5	2360	Term	+	C/S	urinary tract infection	24	29	M	MA	Died at 25 days
6	1950	Term	+	C/S		17	26	M	MA	Died at 10 hr
7	1770	Term	+	V		NA	NA	F	W	Surviving at 6 mo
8	3350	Term	+	C/S	prior drug use	19	NA	M	W	Surviving at 7 mo

+ = Present
– = Absent
M = Male
F = Female
MA = Mexican-American

B = Black
W = White
NA = Not Available
C/S = Cesarian Section
V = Vaginal

TABLE 2. Physical Abnormalities

Case	Encephalocele	Skull and scalp defect	Tissue bands on the face	Abnormal eyes	Cleft lip and/or palate	Limb amputations	Other
1	+	+	+	+	+	+ (U)	
2	+	+	+	+	+	+ (U/L)	
3	+	+	+	+	+	+ (U/L)	coarctation of aorta, omphalocele, simian crease
4	+	+	+	+	+	+ (U/L)	syndactyly (L)
5	+	+	+	+	−	−	high-arched palate, primary hypospadias, clubfoot, simian crease, syndactyly (L)
6	+	+	+	+	+	+ (U)	PDA, primary hypospadias, camptodactyly (U)
7	−	+	+	−	+	+ (U)	
8	−	−	+	+	+	+ (U/L)	clubfoot, simian crease, syndactyly (U/L)

+ = Present − = Absent U = Upper limb L = Lower limb

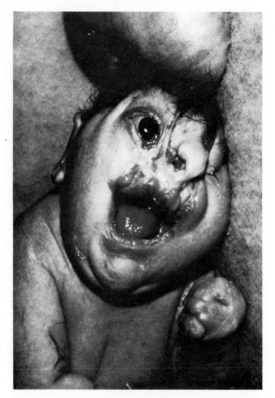

Fig. 1. *Case 1.*

Chromosome examination from blood and fibroblast (skin) cultures by trypsin-Giemsa banding showed an apparently normal 46, XY karyotype.

Although profoundly retarded, this child survived for 25 months. Postmortem examination revealed pneumonia, presumably the cause of death, and the already noted congenital anomalies. No internal congenital anomalies were found. Specifically, the adrenals were normal.

Case 2

A 2960-gm female was born at term by normal vaginal delivery to a 31-year-old Mexican-American woman who was gravida 3 para 3. The pregnancy was unremarkable and the family history was unrevealing. There were 2 normal older sibs. No further pregnancies have occurred.

The infant was found to have multiple, small and asymmetric encephaloceles involving the midfrontal and right parietal skull. The encephaloceles were covered by skin and separated by tissue bands (Fig. 2). Additionally, there was a facial band on the left side of the face extending from the upper lip to the head and another band through the outer canthus of the right eye and inner canthus of the left eye holding the eyes open. The infant had a left cleft lip and palate

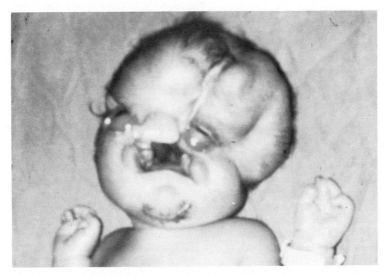

Fig. 2. *Case 2.*

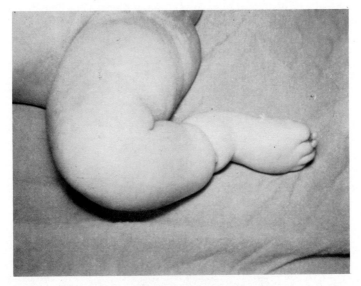

Fig. 3. *Case 2.* Tissue band – leg.

with protrusion of the premaxilla; amputations of the right 3rd and 4th fingers with remaining nubbins of tissue; partially absent left thumb; joining of the distal phalanges of the 2nd, 3rd and 4th left fingers by skin and fibrous tissue; and partially amputated right 1st and 2nd toes. There was a constriction around the right lower leg about 3 cm above the malleolus (Fig. 3).

The child was profoundly retarded and alive at 2½ years of age.

Chromosome analysis from a peripheral blood culture revealed an apparently normal 46, XX karyotype.

Case 3

A 2155-gm male was the second of twins born at term by a normal vaginal delivery to a 17-year-old black primigravida. The first twin, also male, weighed 2700 gm and was normal. Two separate placentas were noted. The mother had severe nausea in the second month and chest pain with weakness of her left leg for several days during the third month of pregnancy. She denied using illicit drugs. Family history was unrevealing. To date the mother has had no further pregnancies.

The infant had multiple asymmetric encephaloceles covered with skin, scalp defects, band-like tissue above the right skull defect and on the right side of the nose, anophthalmia, hypertelorism, left median lid coloboma, a complete bilateral cleft lip and palate, and a wide mouth. An omphalocele, bilateral simian creases and amputations of both upper and lower limbs were also present (Fig. 4). The left hand had distal amputations of more than half of each finger. The right 1st, 2nd and 3rd fingers were missing or represented by small nubbins with band-like constrictions. The left foot had a deep furrow between the 1st and 2nd toes. The right foot had a medially placed great toe; amputations of the 2nd, 3rd and 4th toes; and a small 5th toe with a band-like constriction.

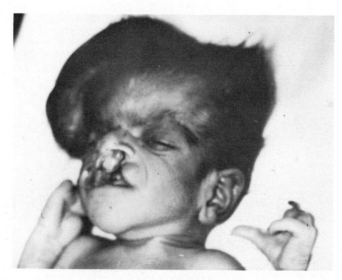

Fig. 4. *Case 3.*

Chromosome analysis from blood using trypsin-Giemsa banding revealed an apparently normal 46, XY karyotype. Blood typing of 14 RBC antigens of each twin was identical.

The affected child died at 1 day of age. At postmortem a coarctation of the aorta distal to the left subclavian artery was found. The adrenal glands were normal.

Case 4

A 2280-gm female was born at term by normal vaginal delivery to a 21-year-old white woman who was gravida 2 para 2.

The mother had ingested LSD and speed (probably amphetamine) in the year prior to conception, and marijuana, hashish and catnip tea early in pregnancy. During her first trimester, she reported the illness and death of her pet hamster and finches, and she had an influenze-like syndrome in the sixth month of pregnancy. She also stated that she was emotionally distraught throughout the pregnancy and changed her residence several times. The maternal half-sib of this child was described at birth as having the eyes stuck together, requiring surgery. This sib was placed for adoption and follow-up was not possible. The rest of the

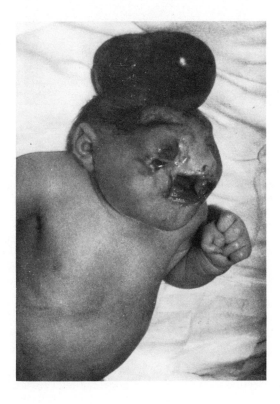

Fig. 5. *Case 4.*

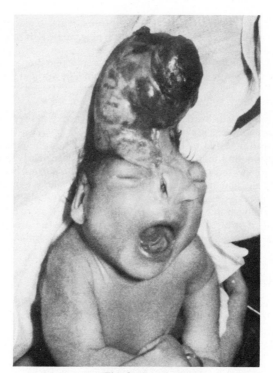

Fig. 6. *Case 5.*

family history was unrevealing. The mother is now pregnant. Ultrasonography and alpha-fetoprotein determination of the amniotic fluid were normal.

The infant had a midline occipital encephalocele with anterior skull defects, facial bands of tissue extending from the lips and the eyelids to the forehead, left anophthalmia, right scalloped eyelid with corneal opacity and misshapen iris and complete bilateral cleft lip and palate (Fig. 5). There were also proximal syndactyly and distal amputations of the right 2nd, 3rd and 4th fingers. There were bifid thumb and 6 digits on the left hand, bilateral syndactyly of the 2nd and 3rd toes and webbing between the 4th and 5th toes. Neurologic examination showed increased muscle tone and absence of deep tendon reflexes and pupillary responses.

Chromosome analysis from blood using trypsin-Giemsa banding showed an apparently normal 46, XX karyotype.

The infant expired at 1 day of age and no autopsy was obtained.

Case 5

A 2360-gm male was born at term to a 24-year-old Mexican-American primigravida by C section for breech presentation and fetal distress. The mother had first trimester bleeding and a probable urinary tract infection. Family history was unrevealing. The mother is now pregnant again.

The anomalies noted at birth included a midline occipital encephalocele with anterior skull defects, a tissue band drawing the right lateral superior ala nasi upward to the encephalocele, hypertelorism, an iris coloboma and a cloudy cornea in the left eye, a folded and notched helix on the left ear, broad and flattened nares, a high-arched palate, first degree hypospadias, bilateral simian creases, bilateral 5th finger clinodactyly, syndactyly of the left 4th and 5th toes, and a right clubfoot (Fig. 6).

Chromosome analysis from blood using trypsin-Giemsa banding showed an apparently normal 46, XY karyotype with a normal variant of No. 9.

The infant was profoundly retarded and died at 25 days of age due to meningitis. No additional anomalies were found at postmortem examination. The adrenal glands were normal.

Case 6

A 1950-gm male was born at term to a 17-year-old Mexican-American primigravida by C section for breech presentation and fetal distress. The mother said that she had episodes of severe abdominal pain during the first month of gestation. Family history was unrevealing. To date the parents have established no further pregnancies.

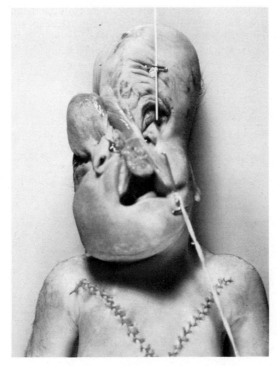

Fig. 7. *Case 6.*

The infant had 2 asymmetric encephaloceles with scalp defects, band-like tissue from the upper lip to the left encephalocele, left microphthalmia, right anophthalmia, malformed nares, bilateral cleft lip and palate, band-like tissue running from the posterior tongue to the hard palate, first degree hypospadias, absent right distal 5th phalanx, hypoplastic nails, camptodactyly of the 2nd through 5th right digits, absent left thumb and a protuberance of band-like tissue over the left radiocarpal area (Fig. 7).

Chromosome analysis from blood using trypsin-Giemsa banding showed an apparently normal 46, XY karyotype with a Gs+ marker, which is considered a normal variant.

The infant expired at 10 hours of age. Additional findings at postmortem examination were agenesis of the brain and a patent ductus arteriosus. The adrenal glands were normal.

Case 7

A 1770-gm female was born at term by normal vaginal delivery to a single white woman who was gravida 2, para 1, ab 1. The attending obstetrician reported an increase in the thickness of the fetal membranes which adhered to the infant's scalp and required manual removal. The pregnancy and family history were unrevealing.

The infant had occipital scalp defects which extended under the left ear and on to the cheek for approximately 5 cm, a small right anterior scalp defect, slight upward slant to the palpebral fissures, left complete cleft palate and gum, indentation on the right upper and left lower lips with excess tissue on the left lip, deformed left helix, flattened nares and amputations of the left 3rd and 4th fingers (Figs. 8 and 9). The infant appeared normal to neurologic examination but required partial gavage feeding.

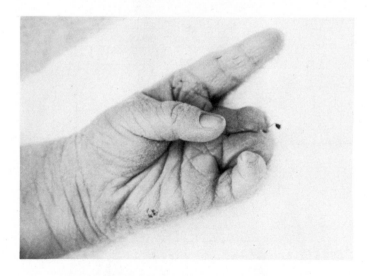

Fig. 8. *Case 7.* Typical digit amputations.

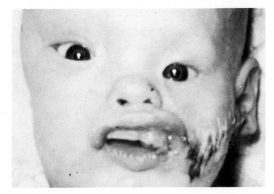

Fig. 9. *Case 7.*

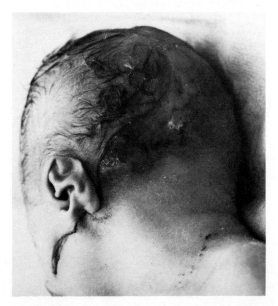

Fig. 10. *Case 7.* Postoperative repair of scalp defect.

Radiographs of the skull showed a defect in the occipital bones. A chest radiograph revealed bilateral pulmonary infiltrates. A pneumoencephalogram was normal. Chromosome analysis from peripheral blood revealed an apparently normal 46, XX karyotype.

The child received a full thickness graft to the occipital lesion before discharge (Fig. 10). No follow-up is available.

Case 8

A 3350-gm male was born at term to a 19-year-old white woman who was gravida 2, para 1, ab 1. A C section was done for cephalopelvic disproportion. The mother had extensive drug history prior to conception, including LSD, heroin, amphetamines and marijuana. She denied drug use during pregnancy. Family history was unrevealing.

The infant had distorting tissue bands of the face; left lid coloboma; complete bilateral cleft lip and palate; syndactyly of the left 1st, 2nd and 3rd fingers; partial syndactyly of the left 4th and 5th fingers; hypoplasia of the distal phalanx of the left 2nd, 3rd and 4th fingers; hypoplasia of the left 5th finger; amputation and tissue band formation at the distal portion of the right finger; webbing between all digits on the right hand; and bilateral simian creases. There were also syndactyly of the left 2nd and 3rd toes and the right 3rd and 4th toes, hypoplasia of all right toes and a right talipes equinovarus deformity.

The facial band deformities consisted of an external band of tissue extending laterally on the left from the cleft lip to the orbit and across the orbit and nares to the right upper forehead, where it fused with a lateral web of tissue from the forehead and the right cleft (Fig. 11). This web partially covered the right eye, and a small cutaneous horn was present in the midforehead.

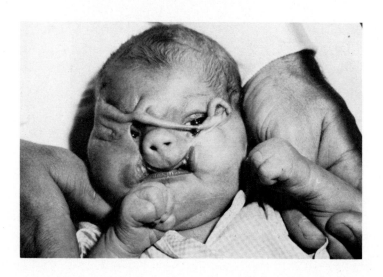

Fig. 11. *Case 8.* Facial deformities.

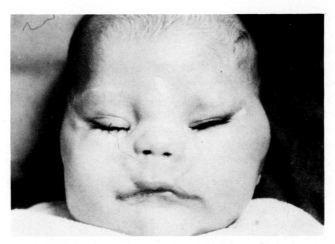

Fig. 12. *Case 8.* Postoperative repair of facial deformities.

Skull radiographs were normal. Chromosome analysis from peripheral blood using trypsin-Giemsa banding revealed an apparently normal 46, XY karyotype with a Gp+ variant.

The infant did well in the nursery, had surgery for the talipes equinovarus and plastic repair of the facial deformities (Fig. 12). At 7 months of age the child is developing normally.

DISCUSSION

Although recognized for many years (2—8), the syndrome of aberrant tissue bands and craniofacial defects has usually been grouped with other malformations (9—11). Recent investigators have attempted to distinguish between this syndrome and other conditions (1, 12, 13).

There are several theories of the pathogenesis of amniotic bands (3—5). Torpin (4) has postulated that amniotic bands are formed when the amnion ruptures early or separates from the chorion. Fetal swallowing of a fibrous band could approximate the scalp to the amnion, hence scalp and skull defects. *Case 7* had the head approximated to the placenta and required removal of the placental tissues from the skull.

Streeter (3) proposed that amniotic bands and the associated malformations were secondary to a developmental abnormality. This proposal is strengthened by the common ectodermal origin of the amnion and the embryonic disk. The malformations in these children, mainly in ectodermal tissue, may have occurred at various time or gestation stages of development, producing the wide variation in deformities. An early defect of the ectoderm could lead to abnormal amnion development, subsequent band formation with amputation or entrapment of fetal parts and possible reabsorption of the parts.

No consistent environmental influence could be identified in these patients. At least 2 of the mothers admitted to drug use, including LSD and marijuana, prior to conception. While there are several reports of infants with limb reduction anomalies born to women who have taken LSD and marijuana (13–16), there is no current consensus that LSD and marijuana are teratogenic (16–19).

Family pedigrees in *Cases 1–8* showed no evidence of increase in birth defects, fetal wastage or consanguinity. The differentiation between this pattern of malformations and anencephaly alone, cleft palate alone or other syndromes (ie Meckel syndrome) (20) is important for genetic counseling. This syndrome of aberrant bands and craniofacial anomalies seems to be sporadic in occurrence and the recurrence risk is considered low.

Physicians should be aware that some of these infants may live a considerable length of time. Two of the severely affected children have survived for more than 2 years. Reconstructive surgery should be considered if the child seems neurologically intact.

SUMMARY

A report of 8 cases of craniofacial anomalies and aberrant tissue bands has been presented. Analyses of the pedigrees, pregnancy histories and chromosome analyses were unrevealing. This syndrome appears to be a sporadic disorder and it is important to delineate it further from other syndromes for which recurrence risks are known.

ACKNOWLEDGMENTS

The authors wish to acknowledge the review of the manuscript by Drs. Miriam G. Wilson and Blanche Bobbitt, and the chromosome analyses by Dr. Joseph W. Towner with the technical assistance of Mr. Paul Brager, Mrs. Talma Dawson, Mr. Paul Nazarian and Mrs. Fay Kaplan.

REFERENCES

1. Jones, K. L., Smith, J. G., Hall, B. D. et al: A pattern of craniofacial and limb defects secondary to aberrant tissue bands. J. Pediatr. 84:90–95, 1973.
2. Rischbieth, H.: Part 4. Hare-lip and cleft palate. In "Treasury of Human Inheritance," Eugenics Laboratory Memoirs XI. London: Dulau & Co., Ltd., 1910, vol. 1, sec. XII[a]
3. Streeter, G. L.: Focal deficiencies in fetal tissues and their relations to intrauterine amputations. Contr. Embryol. Carneg. Instn. 22:1–44, 1930.
4. Torpin, R. C.: "Malformations Caused by Amnion Rupture During Gestation." Springfield: Charles C Thomas, 1968, p. 180.

5. Liban, E. and Abramouici, A.: Fetal membrane adhesions and congenital malformations. Drugs and fetal development. In Klingberg, M. A., Abramouici, A. and Chemke, J. (eds.): "Advances in Experimental Medicine and Biology." New York: Plenum Press, 1972, vol. 27, pp. 337–350.

6. Potter, E.: "Pathology of the Fetus and Newborn." Chicago: Yearbook Medical Publishers, 1972.

7. Baker, C. J. and Rudolph, A. J.: Congenital ring constrictions and intrauterine amputations. Am. J. Dis Child. 121:393–400, 1971.

8. Kohler, H. G.: Congenital transverse defects of limbs and digits. Arch. Dis. Child. 37:263, 1962.

9. Apple, D. J. and Bennett, T. O.: Multiple systematic and ocular malformations associated with maternal L.S.D. usage. Opthalmologica 92:1–303, 1974.

10. Anderson, S. R., Bro-Rasmussen, F. and Tygstrup, I.: Anencephaly related to ocular development and malformations. Am. J. Opthalmol. (Suppl.) 64:559–66, 1967.

11. Woyton, J.: Encephalocele attached to the placenta. Am. J. Obstet. Gynecol. 81:1028–1032, 1961.

12. Chemke, J., Graff, G., Hurwitz, N. and Liban, E.: The amniotic band syndrome. Obstet. Gynecol. 41:332–336, 1973.

13. Hanson, J. W. and Freeman, M. G.: Aberrant tissue bands and multiple congenital defects: An epidemiologic assessment. In Bergsma, D. (ed.): "New Chromosomal and Malformation Syndromes," Birth Defects: Orig. Art. Ser., vol. XI, no. 5. Miami: Symposia Specialists for The National Foundation – March of Dimes, 1975, p. 329.

14. Assemarry, S. R., Neu, R. L. and Gardner, L. I.: Deformities in a child whose mother took L.S.D. Lancet 1:1290, 1970.

15. Blanc, W. A., Mattison, D. R., Kane, R. and Chauhan, P.: L.S.D., intrauterine amputations and amniotic band syndrome. Lancet 2:158, 1971.

16. Carakushansky, G., Neu, R. L. and Gardner, L. I.: Liperide and cannabis as possible teratogens in man. Lancet 2:150–151, 1969.

17. Long, S.: Does L.S.D. induce chromosomal damage and malformations? A review of the literature. Teratology 6:75–90, 1972.

18. Hecht, F., Beals, R. K., Lees, M. H. et al: Lysergic-acid-diethylamide and cannabis as possible teratogens in man. Lancet 2:1087, 1968.

19. Temtamy, S. and McKusick, V. A.: Synopsis of hand malformations with particular emphasis on genetic factors. In Bergsma, D. (ed.): Part III. "Limb Malformations," Birth Defects: Orig. Art. Ser. vol. V, no. 5. White Plains: The National Foundation – March of Dimes, 1969, pp. 125–184.

20. Fried, K., Mundel, G., Reif, A. and Bukovsky, J.: A Meckel-like syndrome? Clin. Genet. 5:46–50, 1974.

Selected Abstracts

THE EFFECT OF MATERNAL RUBELLA ON THE PERMANENT DENTITION

M. M. Cohen and F. E. Gallagher

Developmental Evaluation Clinic, The Children's Hospital Medical Center, Boston, Massachusetts

In a previous report of the effect of maternal rubella on the primary dentition in 38 children with a mean age of 4.2 years, it was found that the incisors and canines in the males were significantly smaller ($p < .05$) than in a normal control group. The females in the rubella group showed no significant size reduction at $p < .05$ level. The 1st maxillary molars in the females were slightly larger when compared to the control group.

The present report is a continuation of the earlier study to observe any late effects on the permanent dentition and includes 16 children with an average age of 7.2 years.

Mesiodistal crown measurement could only be made on permanent teeth that were erupted in the developing dentition. Only permanent incisors and permanent molars were measured. When these measurements were compared with a normal control group, it was found that they were comparable in size.

From this pilot study it appears that there are no late effects on tooth size and the developing dentition in children affected with maternal rubella.

FETAL ALCOHOL SYNDROME WITH NOONAN PHENOTYPE

F. Char

Dept. of Pediatrics, University of Arkansas Medical Center, Little Rock, Arkansas

In 1973, Jones et al described 8 patients with a specific pattern of malformation in the offspring of mothers who were chronic alcoholics. The malformation pattern included short palpebral fissures, epicanthal folds, maxillary hypoplasia with relative prognathism and mild microcephaly. Prenatal growth deficiency

Birth Defects: Original Article Series, Volume XII, Number 5, pages 81–84
© 1976 The National Foundation

and developmental delay were usually present. In 1974, Hall and Orenstein first described an example of the Noonan phenotype in an offspring of an alcoholic mother.

The purpose of this report is to document another case of the fetal alcohol syndrome with Noonan phenotype and to emphasize the importance in differentiating the two conditions.

The patient is an 11-year-old white female, the fifth child of alcoholic parents. Her mother and father were 35 and 42 years of age, respectively, at the time of birth. Her mother died of alcoholic cirrhosis. Two older sibs are living and well. One sib died at 8 days and another was a stillborn. The patient weighed 2466 gm at birth. Developmental landmarks were retarded. At 11 years of age she attends a special school and is doing third grade level work.

Physical examination at age 11 years revealed the following: height 130 cm, weight 25 kg (50th% for 8-year-olds), head circumference 49 cm, and palpebral fissures length 2.3 cm. The metopic suture was prominent. The face was somewhat narrow with slight low-set ears. The mandible was prominent. There was webbing of neck with a posterior trident hairline. The palate was high-arched. The teeth appeared normal and were slightly widely spaced. Intermammary distance was slightly increased. No heart murmurs were heard. There was a wide carrying angle. The clitoris measured 1.5 cm.

Bone age was 9 years with chronologic age of 11 years. Buccal smears showed presence of one Barr body. Chromosome studies obtained from peripheral blood and skin showed 46,XX. Banding techniques failed to show structural defects. Psychometrics showed a verbal IQ of 84, performance IQ of 90 and full scale IQ of 85.

FURTHER OBSERVATIONS ON THE POSSIBLE TERATOGENIC EFFECTS OF SALICYLATES

J. R. McNiel

Kersten Clinic, Fort Dodge, Iowa

Two more cases of the possible teratogenic effect of salicylates (aspirin) in humans are presented. In *Case 1* the mother took 21 mg/kg/day from day 26 to day 32 of gestation. Along with the aspirin, she also regularly drank a soft drink containing 0.36 gm of sodium benzoate (which potentiates the teratogenic effect of aspirin in experimental animals). At birth the infant had a cleft of the upper lip and maxilla. In *Case 2* the mother took 24 mg/kg/day of aspirin for 28 days, from day 39 through day 67 of gestation. She took no other medications in early pregnancy. At birth the infant was anencephalic.

Follow-up of two previous cases, now 5 years old, with deformity of the upper limbs possibly due to the teratogenic effect of salicylates showed no gross abnormalities of mental, emotional or motor development.

CELLULAR RESPIRATION IN ABNORMALLY DEVELOPING EMBRYOS

M. L. Netzloff and O. M. Rennert

Dept. of Pediatrics, University of Florida College of Medicine, Gainesville, Florida

Folic acid antagonists are one class of drugs shown to be definitely teratogenic in man. Use of one such compound, 9-methyl·pteroylglutamic acid (9-methyl PGA) in rats on days 10, 11 and 12 of pregnancy results in: 1) a 95% incidence of congenital malformations in offspring at term and 2) altered lactate dehydrogenase (LDH) isozymes in embryonic tissues. The latter observation involved LDH-5, which is an allosteric regulatory enzyme for anaerobic/ aerobic metabolism. To study embryonic oxidative metabolism, pregnant Long-Evans rats were subjected to the same teratogenic regimen of folic acid restriction and antagonism used in the preceeding experiments. Cells were dispersed on day 13 by passing the embryo twice through a 5 ml Mohr pipette. Room air was used with the day 13 dispersed and day 11 intact embryos, and 100% oxygen with day 12 intact embryos. Oxygen consumptions were measured using the direct Warburg and oxygen electrode techniques. Use of larger intact embryos in room air, and on day 13 even in 100% oxygen, is limited by restriction to oxygen diffusion imposed by thick layers of tissue. The use of dispersed embryonic cells on day 13 or 100% oxygen on day 12 obviates these difficulties. The means ± standard errors for day 13 experimental and control respiratory rates are 12.3 ± 0.7 and 17.9 ± 0.5 μl/hr/mg protein, respectively, and differ at a 99% level of significance. Similar day 12 experimental and control data are 18.5 ±1.0 and 18.8 ± 1.0, respectively. Preliminary day 11 data also show no statistically significant difference between mean experimental and control uptakes. Because a day 13 embryo is distinct both developmentally and in duration of treatment from embryos earlier in gestation, the experimental and control oxygen uptakes differ on this day, but not on days 11 and 12. This finding correlates well with the observation that treatment with the 9-methyl PGA regimen for less than 72 hours results in far fewer malformations and resorptions than if the full 72-hour period is used.

BIRTH DEFECTS MONITORING PROGRAM

L. D. Edmonds, L. M. James, J. E. Nissim and J. W. Flynt, Jr.

Birth Defects Branch, Center for Disease Control, Atlanta, Georgia

The Birth Defects Monitoring Program, a national program to monitor and analyze hospital discharge data on congenital malformations, became operational in December 1974. The BDMP is a cooperative program of the Commission on Professional and Hospital Activities (CPHA), The National Foundation-March of Dimes, the National Institute of Child Health and Human Development and the Center for Disease Control (CDC). Over 1 million annual births, comprising 33% of the U.S. total, are systematically monitored for increases, trends and unusual patterns of birth defect incidence. Approximately 80% of CPHA's PAS hospitals with obstetric services have granted use of their data. The BDMP provides for computer comparisons of observed and expected numbers of cases for some 200 defect categories to detect statistically significant increases.

Early evaluation of BDMP data shows some notable patterns of occurrence. Among the earliest noted increases detected by the monitoring system is patent ductus arteriosus. Incidence rates per 10,000 total births for patent ductus arteriosus in the United States have steadily increased from 3.5 in 1970 to 7.7 in 1974. Rates for the 1970–1974 period are highest in the Mountain (9.2) and Pacific (8.8) Census Divisions and lowest in East South Central (2.4) and the New England (3.6) Census Divisions. Black rates (8.8) are twice those for whites (4.4) and black females have the highest sex-race specific rates.

Comparisons of defect incidence have been made between BDMP and other data sources as one way of assessing the validity of the monitored data. Spina bifida has been well described in the literature and reported as having higher mortality rates in the Eastern United States than in the West. BDMP data for the total 1970–1973 time period show similar geographic differences with the lowest reported incidence rates (per 10,000 total births) in the Mountain (5.3) and Pacific (6.2) Census Divisions and the highest rates in the East South Central (10.8) and South Atlantic (9.5) Census Divisions. BDMP data are also consistent with earlier studies showing higher rates among whites than blacks (8.0 vs 4.7 per 10,000 total births).

SECTION II:
CYTOGENETIC DISORDERS

Usefulness of Chromosome Catalog in Delineating New Syndromes

Digamber S. Borgaonkar, PhD,* Yves E. Lacassie, MD†
and Claude Stoll, MD‡

Clinical genetics has established itself as a specialty of both medicine and genetics in the last decade or so. Its establishment differed significantly from other branches of medicine in that it developed from a basic science into a clinical specialty (1). Clinical cytogenetics is, of course, a more limited subspecialty of clinical genetics. On clinical grounds one suspects that a patient has a chromosomal abnormality not because the patient fits a particular syndrome, but because the aggregation of factors points towards a chromosomal basis. These factors often include low birthweight, mental retardation and multiple congenital abnormalities. When such a patient is found to have an abnormal karyotype the next difficult question is: Is this the basic defect in the patient? Usually it is answered by comparing the clinical features of the new patient to other patients with similar abnormal karyotypes. Whenever similarities are found in the phenotype, an attempt is made to describe a new syndrome and to collect more similar cases to justify the delineation of this new syndrome. These have been the usual steps since the early years of human cytogenetics when syndromes such as Patau, Edwards, and cri-du-chat were established, and the basic defects of previously delineated conditions such as Down, Klinefelter and Turner were defined.

In the past 5 years, with the advent of the banding techniques, it became possible to identify segments of all 24 human chromosomes. Therefore, it is conceivable that variations in all of these regions, if they exist, could be described.

*Supported in part by USPHS grants GM 19489 and GM 00795.
†Supported in part by grant PA71-468A from the Ford Foundation.
‡Supported in part by a grant from Eli Lilly Company

Birth Defects: Original Article Series, Volume XII, Number 5, pages 87–95
© 1976 The National Foundation

We have shown elsewhere (2) that all of the autosomes are involved in inter-changes which are compatible with live birth. With *Chromosomal Variation in Man: A Catalog of Chromosomal Variants and Anomalies* (3), a publication we recently compiled and will continue updating, it is possible to identify easily reports with similar changes in the karyotype at every band. Two or more reports of a similar abnormality of partial monosomy or trisomy can thus be traced through the catalog. If there are similarities in clinical descriptions in at least 2 different cases of such reports, then an attempt could be made to delineate the condition. We may find that many karyotypic changes have manifestations which, though similar, are not distinctive enough to delineate a syndrome (eg 20r, as will be shown later in this report), or otherwise there are individuals with specific abnormal karyotypes whose phenotypes are too divergent (eg 47,XYY) (4).

The well-recognized chromosomal variations and chromosomal syndromes are listed in Table 1. Most of these conditions were described in the 1960s but some (eg 8+, 9+, centromere region of chromosome 3) were described in the 1970s. It is conceivable that this list will be outdated by the time this manuscript is published.

Some chromosomal syndromes which have been described rather recently are listed in Table 2. Various authors made these proposals on the basis of their own case material and the data from the literature. Usually the literature review was attempted by scanning various indexes (eg "Index Medicus," Excerpta Medica: "Human Genetics Abstracts") and/or by obtaining listings from "MEDLINE" or "MEDLARS" and/or by scanning a few specialty journals for the past few years.

TABLE 1. Well-Recognized Chromosomal Variations and Syndromes

Chromosomal Syndromes	Chromosomal Variations
4p− Wolf-Hirschhorn	1qh
4p+	3 centromere region
5p− Cri-du-chat	4 centromere region
8+	9qh
9+	13p
9p+ Rethore	14p
13+ Patau	15p
13q− and 13r	16qh
18+ Edwards	21p
18p−	22p
18q−	Yq
18r(18p− and 18q−)	
21+ Down	
22+	
21q− and 21r antimongolism	
22q− and 22r	
X and Y excluded	

TABLE 2. Recently Recognized New Chromosomal Syndromes

Chromosomal Region	Selected Reference
5p14 to 5pter, trisomy	Stoll et al (10)
7q+	Berger et al (11)
9p−, monosomy	Alfi et al (6)
10p+, trisomy*	Schleirmacher et al (14)
10q24 to 10q26, trisomy	Yunis and Sanchez (16)
13q+ proximal segment, trisomy	Escobar et al (23)
13q+ distal segment, trisomy	Escobar et al (23)

*See also Ref. 15.

This procedure was rather time-consuming and was not entirely satisfactory due to the inherent time lag between the finding of an abnormal karyotype in a patient, subsequent work-up, submission of the manuscript and its publication. Moreover, this procedure was independently followed by every investigator who attempted to describe a new chromosomal condition.

In Table 2 are listed at least 3 conditions (9p− and 13q+ proximal and distal segment trisomies) on which additional data were presented at this conference (5–7). It is likely that some of these syndromes may just have attained enough stature to justify their inclusion in Table 1.

METHOD

In the *Catalog* (3) we have included unique and rare chromosomal variations and anomalies by chromosome, arm, region, band or subband. Therefore, it was convenient to look for new chromosomal variants or syndromes. The procedure followed was as such: All reports for each chromosome were evaluated for variant regions, for monosomy or trisomy for the entire chromosome or a segment of it. If there were 2 or more cases (ideally in 2 unrelated families) of a similar monosomy or trisomy for any chromosomal region, then the clinical details on the patients were compared with one another. In some instances it was necessary to obtain additional clinical details on the case by personal correspondence. If it was found that a case could be made for a new syndrome, then an effort was made in this direction. In Table 3 we suggest additions to the lists of chromosomal variants and syndromes arrived at by this method. Only more recent and representative references are provided here, since a detailed listing of all cases can be found in the *Catalog.*

RESULTS

Specific characteristics of each new syndrome follow:

2p22 to 2pter, Trisomy

Face: antimongoloid palpebral slants, pocket-shaped inferior eyelid, long

TABLE 3. Suggested Additions to the Lists of Chromosomal Variants and Syndromes

Chromosome Region	Reference
2p22 to 2pter, trisomy	Stoll et al (8)
2q, variant region	Several authors*
3p21 to 3p26, trisomy	Rethore et al (9)
11p+	Falk et al (20)
	Sanchez et al (19)
	Thompson (18)
11q+	Rott et al (22)
	Dinno et al (21)
14q, partial trisomy	Reiss et al (27)
	Short et al (26)
	Fawcett et al (28)
15q, partial trisomy	Crandall et al (29)
	Fujimoto et al (30)
	Watson and Gordon (31)
17p, variant region	Several authors*
20r	Atkins et al (32)
	Faed et al (33)
	Uchida and Lin (34)

*Detailed references can be found in Borgaonkar's *Catalog* (3).

philtrum, hypertelorism, epicanthus; nose: triangular nostrils, large subcloison, duplication of the extremity (bifid tip), root regular hyperbole shaped; articular nodosities (knots) (8).

3p21 to 3p26, Trisomy

Head: microbrachycephaly; forehead: prominent, high; face: square, hypertelorism, epicanthus, microstomia with falling corners of the lips, retrognathism, short neck; dermatoglyphics: numerous loops on fingers; cardiopathy; intestinal (GI) abnormalities (9).

5p14 to 5pter, Trisomy

Chin: prominent, large, shoe-shaped and well-developed, very developed mandibular angle; normal head circumference (10).

7q+ Trisomy

There are 2 types; one for the trisomy of the region 7q21 or 22 to 7q31 and the other for region 7q31 to 7qter. The characteristics for both these conditions are presented in Table 4 (11).

9p−, Monosomy

Face: trigonocephaly, prominent forehead, wide flat nasal bridge, anteverted nostrils, long upper lip, low-set ears, mongoloid slant of the eyes, epicanthal fold,

TABLE 4. Characteristics of 7p Partial Trisomies*

	Trisomy for Regions	
	7q31 to 7qter	7q21 or 22 to 7q31
Low birthweight	+	−
Delayed stature development	+	−
Prominent occiput or forehead	+	+
Microcephaly	−	+
Palpebral slants	horizontal narrow	narrow
Epicanthus	−	+
Nose root	normal	flat
Retrognathism	+	−
Micrognathism	+	−
Cleft palate	+	−
Hypotonia	±	+
Ears	low-set	large

*Constructed mainly from data in Berger et al (11).
+ = present
− = absent

hypertelorism, variable micrognathia; short neck; wide-set nipples; long fingers, mostly secondary to long middle phalanges; predominance of whorls on fingers. Congenital heart disease and congenital hernias are sometimes present (12). It may be that this syndrome ought to be moved to the category of well-established syndromes in Table 1 (5, 6).

10p11 to 10pter, Trisomy

Severe mental retardation; delayed growth since intrauterine life; dolichocephaly; sutures and fontanel wide open at birth and early infancy, high and bulky forehead; dysplastic and low-set ears; abnormal palate, usually cleft; hypoplastic and hypotonic skeleton muscles with flexion joint abnormalities of limbs, abnormal palmar flexion creases, hypoplasia of phalanges with clinodactyly, camptodactyly; GU anomalies (usually renal); occasionally congenital heart defects (aortic coarctation and pulmonary stenosis); ocular coloboma (13–15).

10q24 to 10q26, Trisomy

Severe mental retardation and delayed growth; microcephaly, flat and round face, spacious forehead, arched and wide-set eyebrows, blepharophimosis, ptosis of eyelids, antimongoloid slants, microphthalmia, small nose and/or depressed nasal bridge, long philtrum and prominent upper lip; micrognathia, and malformed and/or low-set ears; short neck; webbing of fingers or toes; deep plantar furrows; occasionally a weak cry and a simian crease are present (16, 17).

11p, Trisomy

Mental retardation; short stature; unusual face; prominent frontal bossing, enophthalmus, strabismus, nystagmus, wide nasal bridge, abnormal and/or low-set ears; hypogonadism; urinary tract anomalies; abnormalities of hands and feet; simian creases (18—20).

11q, Trisomy

Low birthweight; mental retardation; delayed growth; broad nose; micrognathism; abnormal and low-set ears; cleft palate; agenesis corpus callosum; hypogonadism; urinary system malformations; congenital heart defects (21, 22).

13q, Trisomy

The features present in both of the trisomies of proximal and distal segments are those suggestive of 13+ trisomies. However, a more benign clinical course and longer life-span are suggestive of partial trisomy. Table 5 lists the phenotypic characteristics (23—25).

14q, Trisomy

Shortened stature since birth; microcephaly; hypotelorism; microphthalmia or small palpebral fissures; broad nose; micrognathia; low-set and abnormal ears; camptodactyly and other minor skeletal abnormalities. Also hypertonia, congenital heart disease, and cardiovascular and respiratory abnormalities have been reported (26—28).

TABLE 5. Phenotypic Characteristics of 13q+ Partial Trisomies*

13q+ Distal Segment	13q+ Proximal Segment
Respiratory distress postpartum†	
Bossing of the forehead‡	Other nonspecific signs
Low-set ears†	
Capillary hemangiomata†	
Polydactyly†	
Feet deformity†	

	Absent	Severe growth retardation Arrhinencephaly Microphthalmia Coloboma Cleft lip/palate Micrognathia Hyperconvex fingernails	Present

*Constructed mainly from data in Escobar et al (23, 25).
†Same as 13+
‡Opposite to 13+

15q, Trisomy

Mental retardation; hyperactivity; epilepsy; strabismus, antimongoloid slant; abnormally set ears; spindle fingers; minor facial and limb anomalies (29–31).

20r

Neurologic impairment including EEG changes; low IQ; behavioral and psychologic problems since infancy or childhood (32–34).

COMMENTS

We have shown that the *Catalog* can be utilized quite profitably in the procedure of delineation of new chromosomal syndromes. It is essential, however, to be cautious in this approach, since additional details may need to be gathered by further correspondence (15) with the authors of reports. This is in part due to the practice, which incidentally most of us follow, of describing only those features which at the time of the original report may seem pertinent. There are instances where a particular pattern or effect may become evident during the examination of case material — such as involvement of chromosome 16 in many types of aberrations with invariable association of miscarriages and abortion — but where delineation of a syndrome is not possible. Similarly, large families with inversions of chromosomes 3, 9 and 10 are being described and it would be of interest to compare cases with similar duplications and deficiencies of chromosomal material. It is also possible that a much more fruitful exercise would be to utilize the data from the *Repository of Chromosomal Variants and Anomalies* (35) for the purposes that are mentioned above. Since a large majority of the data in the *Repository* (that we are assembling) is unpublished material from the contributing laboratories, it should be more rewarding than the present study.

It is suggested that trisomies of 2p, 3p, 11p, 11q, 14q and 15q regions of chromosomes are likely candidates for a syndrome. Since the *Catalog* includes not only most of the published papers but also a few significant personal communications, a later review of these data will be helpful in the evaluations.

REFERENCES

1. McKusick, V. A.: The growth and development of human genetics as a clinical discipline. Am. J. Hum. Genet. 27:261, 1975.
2. Borgaonkar, D. S.: Autosomal abnormalities and the banding techniques. In Emery, A. E. H. (ed.): "Modern Trends in Human Genetics." London: Butterworth, Ltd., 1975, vol. 2, pp. 1–30.
3. Borgaonkar, D. S.: "Chromosomal Variation in Man; A Catalog of Chromosomal Variants and Anomalies." Baltimore: The Johns Hopkins University Press, 1975.
4. Borgaonkar, D. S. and Shah, S. A.: The XYY chromosome male — or syndrome? Prog. Med. Genet. 10:135, 1974.

5. Allderdice, P. W., Heneghan, W. D. and Felismino, E. T.: 9pter→p22 deletion syndrome: A case report. This volume.
6. Alfi, O. S., Donnell, G. N. and Derencsenyi, A.: The 9p– syndrome. This volume.
7. Wilroy, R. S., Summitt, R. L., Martens, P. et al: Partial monosomy and partial trisomy for different segments of chromosome 13 in several individuals of the same family. This volume.
8. Stoll, C., Messer, J. and Vors, J.: Balanced t(2;14) in a mother and partial trisomy 2p in two children. Ann. Genet. (Paris) 17:193, 1974.
9. Rethore, M. O., Lejeune, J., Carpentier, S. et al: Trisomy for the distal portion of the short arm of chromosome No. 3 in three sibs. First instance of chromosomal insertion: Ins(7;3) (q31;p21p26). Ann. Genet. (Paris) 15:159, 1972.
10. Stoll, C., Rethore, M. O., Laurent, C. and Lejeune, J.: Reverse type of the cri-du-chat syndrome: Trisomy 5p. Arch. Fr. Pediatr. 32:551, 1975.
11. Berger, R., Derre, J. and Ortiz, M. A.: Partial trisomy for the long arm of number 7 chromosome; a case (bands 7q22 and 7q31) Nouv. Presse Med. 3:1801, 1974.
12. Alfi, O. S., Donnell, G. N., Crandall, B. F. et al: Deletion of the short arm of chromosome No. 9 (46,9p-). A new deletion syndrome. Ann. Genet. (Paris) 16:17, 1973.
13. Hustinx, T. W. J., ter Haar, B. G. A., Scheres, J. M. J. C. and Rutten, F. J.: Trisomy for the short arm of chromosome No. 10. Clin. Genet. 6:408, 1974.
14. Schleirmacher, E., Schliebitz, U., Steffens, C. et al: Brother and sister with trisomy 10p: A new syndrome. Humangenetik 23:163, 1974.
15. Cantú, J. -M., Salamanca, F., Buentello, L. et al: The 10p trisomy. A report of two cases due to a familial translocation rcp(10:21) (p11;p11). Ann. Genet. (Paris) (In press.)
16. Yunis, J. J. and Sanchez, O.: A new syndrome resulting from partial trisomy for the distal third of the long arm of chromosome 10. J. Pediatr. 84:567, 1974.
17. Dutrillaux, B., Laurent, C., Robert, J. M. and Lejeune, J.: Inversion pericentrique, inv(10), chez la mere et aneusomie de recombination, inv(10), rec(10), chez son fils. Cytogenet. Cell Genet. 12:245, 1973.
18. Thompson, H.: Familial chromosomal translocation with distinctive phenotype due to effective trisomy, No. 9p. In Bergsma, D. (ed.): "New Chromosomal and Malformation Syndromes," Birth Defects: Orig. Art. Ser., vol. XI, no. 5. Miami: Symposia Specialists for The National Foundation – March of Dimes, 1975, pp. 213–216.
19. Sanchez, O., Yunis, J. J. and Escobar, J. I.: Partial trisomy 11 in a child resulting from a complex material rearrangement of chromosomes 11, 12 and 13. Humangenetik 22:59,1974.
20. Falk, R. E., Carrel, R. E., Valente, M. et al: Partial trisomy of chromosome 11: A case report. Am. J. Ment. Defic. 77:383, 1973.
21. Dinno, N. D., Silvey, G. L. and Weisskopf, B.: 47,XY,t(9p+; 11q+) in a male infant with multiple malformations. Clin. Genet. 6:125, 1974.
22. Rott, H. D., Schwanitz, G., Grosse, K. P. and Alexandrow, G.: C11/D13 translocation in four generations. Humangenetik 14:300, 1972.
23. Escobar, J. I., Sanchez, O. and Yunis, J. J.: Trisomy for the distal segment of chromosome 13. A new syndrome. Am. J. Dis. Child. 128:217, 1974a.
24. Wilroy, R. S., Summitt, R. L. and Martens, P. R.: Partial trisomy for different segments of chromosome 13 in several individuals of the same family. In "New Chromosomal and Malformation Syndromes," op. cit., pp. 217–222.
25. Escobar, J. I. and Yunis, J. J.: Trisomy for the proximal segment of the long arm of chromosome 13. A new entity? Am. J. Dis. Child. 128:221, 1974b.
26. Short, E. M., Solitaire, G. B. and Breg, W. R.: A case of partial 14 trisomy 47,XY(14q–)+ and translocation t(9p+, 14q–) in mother and brother. J. Med. Genet. 9:367, 1972.

27. Reiss, J. A., Wyandt, H. E., Magenis, R. E. and Hecht, F.: Mosaicism with translocation: Autoradiographic and fluorescent studies of an inherited reciprocal translocation t(2q+, 14q−). J. Med. Genet. 9:280, 1972.

28. Fawcett, W. A., McCord, W. K. and Francke, U.: Trisomy 14q−. In "New Chromosomal and Malformation Syndromes," op. cit., pp. 223−228.

29. Crandall, B. F., Muller, H. M. and Bass, H. N.: Partial trisomy of chromosome number 15 identified by trypsin-Giemsa banding. Am. J. Ment. Defic. 77:571, 1973.

30. Fujimoto, A., Towner, J. W., Ebbin, A. J. et al: Inherited partial duplication of chromosome No. 15. J. Med. Genet. 11:287, 1974.

31. Watson, E. J. and Gordon, R. R.: Case reports: A case of partial trisomy 15. J. Med. Genet. 11:400, 1974.

32. Atkins, L., Miller, W. L. and Salam, M.: A ring-20 chromosome. J. Med. Genet. 9:377, 1972.

33. Faed, M., Morton, H. G. and Robertson, J.: Ring F chromosome mosaicism (46,XY, 20r/46,XY) in an epileptic child without apparent hematological disease. J. Med. Genet. 9:470, 1972.

34. Uchida, I. A. and Lin, C. C.: Ring formation of chromosomes Nos. 19 and 20. Cytogenet. Cell Genet. 11:208, 1972.

35. Borgaonkar, D. S., Bolling, D. R., Blair, S. and Classon, R.: "International Registry of Abnormal Karyotypes. Provisional Listing." Baltimore: The Johns Hopkins University School of Medicine, May, 1975; and "Repository of Chromosomal Variants and Anomalies," Baltimore: The Johns Hopkins University School of Medicine, August 1975.

E Trisomy Phenotype Associated With Small Metacentric Chromosome and a Familial Y-22 Translocation*

Kenneth W. Dumars, MD , Gayle Fialko and Eunice Larson, MD

A review of the literature reveals a number of case reports often entitled "multiple congenital malformations associated with a small metacentric chromosome." The small chromosome is variously described as metacentric, satellited, bisatellited, and often in satellite association. This case report and review leads us to postulate and delineate several *clinical* entities associated with a small and additional chromosome.

CASE REPORT

A male newborn infant was transferred to the Neonatal Intensive Care Unit because of respiratory distress. The child's birthweight at 38 weeks' gestation was 1550 gm, length 41 cm and OF 29.25 cm. This is the first pregnancy for a 47-year-old mother and a 51-year-old father. The delivery was complicated by a placenta previa. At birth the following abnormalities were noted: bulky occiput, antimongoloid slant to eyes, retrognathia, clinodactyly, 5th finger overlapping 4th, systolic murmur consistent with septal defect or AV canal variant, simian line, small penis, undescended testes and hypotonia (Fig. 1). The infant developed generalized seizures and respiratory distress requiring assisted ventilation.

Pertinent laboratory work included normal hct, Hb and chemistries, but 15,000 WBC with 26–40% eosinophils were found upon serial determinations. The infant is A positive; mother O negative, Coombs' negative and ABO positive. Radiographs revealed a small cranial vault and multiple anomalies of upper thoracic vertebras and ribs. Dermatoglyphics revealed hypoplastic dermal ridges, but where visible, a predominance of volar arches. A buccal smear was X-chromatin negative; however, bipartite Y chromatin was found in a high percentage of the cells obtained from a buccal smear.

A family pedigree (Fig. 2) revealed this child to be the first infant born to this mother. Additional studies on the father's children from his first marriage have not been completed due to their living in South America.

*Supported in part by a grant obtained from The National Foundation — March of Dimes.

Birth Defects: Original Article Series, Volume XII, Number 5, pages 97–104

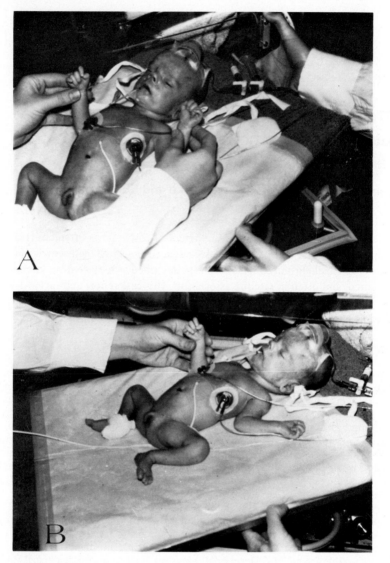

Fig. 1. A and B) Proband. Note the position of fingers, the bulky occiput and retrognathia.

The infant died at 26 days of age after a course of unremitting respiratory distress and generalized seizures. Postmortem examination revealed prematurity consistent with 32–33 weeks' gestation and the following multiple congenital anomalies: infundibular stenosis, pulmonary outflow tract; ventricular septal defect, small; left ventricular hypertrophy; polyscystic right kidney (Potter

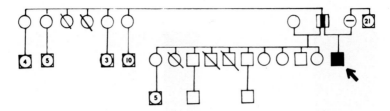

Fig. 2. Family pedigree.

type II); vertebral body anomalies with hemivertebras and butterfly vertebras, upper thoracic region; broad sternum with abnormal growth center; 13 ribs, left; 11 ribs, right; Meckel diverticulum; and adrenal hypoplasia, bilateral (0.4/5.0 gm). Also noted were: hyaline membrane disease, resolving; acute interstitial pneumonitis; subarachnoid hemorrhage, old bilateral, dorsal surface of cerebellum; cerebral edema, moderate; extramedullary hematopoiesis, liver, spleen, lung, and stomach; and pancreatic islet cell hyperplasia and hypertrophy.

Cytogenetic Study

Forty-seven chromosomes were found in cells obtained from skin and lymphocyte cultures. The extra chromosome in metaphase resembled a small metacentric chromosome, but when seen in late prophase it appeared more submetacentric (Fig. 3). No satellites were seen; however, association with acrocentric chromosomes was seen in 35% of metaphase figures analyzed. (The presence of satellites does not exclude origin from E group, since satellites have been described on E17–18.)

An additional finding was the presence of bipartite brilliant fluorescence in the interphase nucleus. Q- and C-banding of the chromosome karyotype confirmed this, ie the long arm (q) of the Y and the small arm (p) of one G22 stained brilliantly with quinacrine (Fig. 3). The same abnormality was found in the father's karyotype (Fig. 4). The mother's karyotype was normal (46,XX).

DISCUSSION

Y-autosome (1–3) and Y to X translocations (4) have previously been described. In no previous case report, nor in this case, is it believed that the additional heterochromatic segment of the Y is responsible for the production of these phenotypic alterations. In this case the Y-22 translocation is a marker chromosome also present in the father.

A review of the literature prompted us to separate several clinical entities (Table 1) associated with an additional small metacentric chromosome, presumably partial deletions of the E- and D-group chromosomes. Deletions of additional G chromosome(s), particularly the cat-eye syndrome as described by numerous investigators (5–7), will not be considered. Additional syndromes which seem to be emerging are as follows:

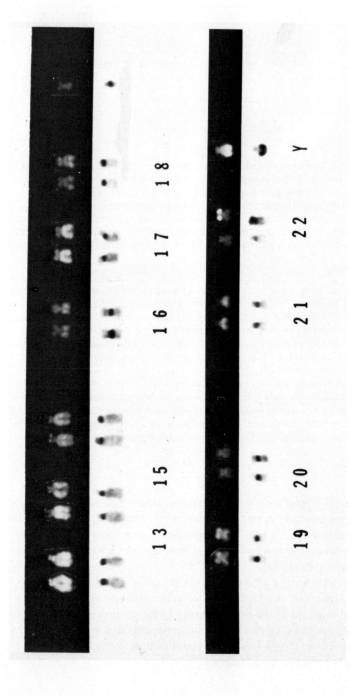

Fig. 3. A partial karyotype of the proband revealing the intense normal fluorescence of the Y chromosome with equally intense fluorescence of the small arms (p) of a G22. This was confirmed by C-banding.

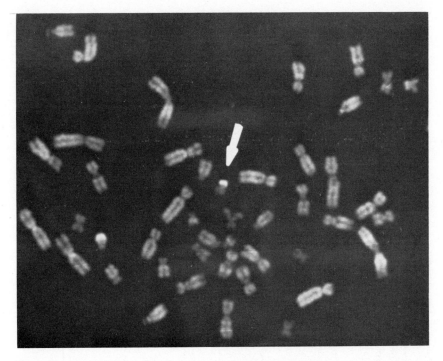

Fig. 4. Karyotype of the proband's father revealing intense fluorescence of the small arm (p) of a G22 (arrow) as well as the normally fluorescent Y chromosome.

1) Infants, as described in this report and by others (8, 9), who present with a phenotype similar to the E18 trisomy and are associated with a small metacentric chromosome without satellites. This extra chromosome is not found in parents or normal sibs. In our case the late prophase figures are not as meta-centric and suggest a deleted long arm of E17 or E18; the latter, in our opinion, being more likely. However, this belief is based upon clinical similarity to the E18 trisomy rather than specific chromosome identification. As stated by Hamerton (10), there is no direct evidence this clinical picture is due to an ab-normal chromosome from the E group.

2) A group of 5 patients described independently by an equal number of in-vestigators (11–15). This group of patients was reported at 8–11 years of age and and all presented with moderately severe mental retardation, microcephaly, facial asymmetry, scoliosis and usually with cerebral palsy. All were below the 10th percentile for height and weight. This phenotype was associated with a meta-

TABLE 1. Phenotypes Associated With an Additional Metacentric Chromosome Originating From the D or E Group

No. Cases	Phenotype Proband	Presence and Characteristics of Metacentric Chromosome				Authors
		Proband	Parents and Sibs	Satellites	Satellite Association	
3	Similar to E trisomy	+	0	0	±	Frøland et al (8) / Gustavson et al (9) / Present case
5	Mental retardation Microcephaly Facial asymmetry Scoliosis Cerebral palsy	+	0	0	0	Mukherjee et al (11) / Ishmael and Laurence (12) / Ferrante et al (13) / Armendares et al (14) / Finley et al (15)
12	Inconstant phenotype	+	+	±	±	Borgaonkar et al (18) / Tamburro and Johnson (19) / Taft et al (20) / Mukherjee and Burdette (21) / Pfeiffer et al (23) / Hoehn et al (24) / Abbo and Zellweger (25) / Tangheroni et al (26) / Crandall et al (27)
3	Normal	+	±	±	±	Smith et al (16) / Walzer et al (17) / Borgaonkar et al (18)

+ = Consistently present
± = Not consistently present
0 = Not present

centric chromosome manifesting inconstant labeling with tritiated thymidine. The mass of the deleted chromosome led the investigators to postulate its origin from the small arm (p) of E16 or 17.

3) A third group was identified in screening programs of phenotypically normal individuals (16—18). There was found an additional small bisatellited chromosome, postulated to be due to centric fusion of acrocentric chromosomes presumably of the D group.

There are still a number of patients (19—27) with small metacentric chromosomes found in the proband but with an inconstant phenotype. In addition, the additional small chromosome was found in a number of unaffected parents and sibs of the proband.

It is tempting to postulate a partial trisomy of short arm of E18 in the first group and a partial E16 or E17 in the second. The bisatellited small chromosome in persons with a normal phenotype (group 3) is consistent with centric fusion of the D or G chromosomes. However, in keeping with David Smith's (20) admonition 10 years ago in his commentary following the presentation of a similar patient, caution must be observed and more evidence generated for this hypothesis.

SUMMARY

As a result of this case report, several entities are postulated due to an extra metacentric D or E chromosome: 1) infants presenting with a phenotype similar to the E18 trisomy; however, the karyotype can be interpreted as either a deleted E or D chromosome; 2) another group of children all presenting with mental retardation, facial asymmetry, scoliosis and cerebral palsy, postulated due to a partial trisomy of E16 or E17; 3) individuals with a normal phenotype, but chromosomally presented with an additional satellited metacentric chromosome consistent with centric fusion of a D or G chromosome and 4) children presenting with an inconsistent phenotype and chromosomally presenting with an extra chromosome manifesting satellites or satellite association; the same chromosome abnormality often is found in unaffected parents and/or sibs.

REFERENCES

1. Reitalu, J.: A familial Y-22 translocation in man. Hereditas 74:155—160, 1973.
2. Lundsteen, C. and Philip, J.: Y-22 translocation in a YY male. Cytogenet. Cell Genet. 12:53—59, 1973.
3. Friedrich, U. and Nielsen, J.: Presumptive Y-15 and Y-22 translocation in two families. Hereditas 71:339—342, 1972.
4. Khudr, G., Benirschke, K., Judd, H. L. and Strauss, J.: Y to X translocations in a woman with reproductive failure. JAMA 226:544—549, 1973.
5. Weber, F. M., Dooley, R. R. and Sparkes, R. S.: Anal atresia, eye anomalies, and an additional small abnormal acrocentric chromosome (47XX, Mart): Report of a case. J. Pediatr. 76:594—597, 1970.
6. Bühler, E. M., Mehes, K., Muller, H. and Stalder, G. R.: Cat-eye syndrome, a partial trisomy 22. Humangenetik 15:150—162, 1972.

7. Gerald, P. S., Davis, D., Say, B. M. and Wilkins, J. L.: A novel chromosomal basis for imperforate anus (the "cat's-eye" syndrome). Program and abstracts 78th annual meeting, "The American Pediatric Society," 1968.
8. Frøland, A., Holst, G. and Terslev, E.: Multiple anomalies associated with an extra small autosome. Cytogenet. Cell Genet. 2:99–106, 1963.
9. Gustavson, K. H., Atkins, L. and Patricks, I.: Diverse chromosomal anomalies in two siblings. Acta Paediatr. (Uppsala) 53:371, 1964.
10. Hamerton, J. L.: "Human Cytogenetics." New York: Academic Press, 1971, vol. II, pp. 365–370.
11. Mukherjee, A. B., Partington, M. W., Simpson, N. E. and Walmsley, K. A.: Multiple anomalies associated with a small extra metacentric autosome. J. Med. Genet. 5:329–334, 1968.
12. Ishmael, J. and Laurence, K. M.: An extra small metacentric chromosome in a mentally retarded boy. J. Med. Genet. 5:335–340, 1968.
13. Ferrante, E., Bruni, L., Laurenti, F. and Coglioti, G.: Parziale trisomia autosomica in soggetto affetto da ritardo mentale e lievi note dismorfiche. Minerva Pediatr. 20:522–529, 1968.
14. Armendares, S., Buentello, L. and Salamanca, F.: An extra small metacentric autosome in a mentally retarded boy with multiple malformations. J. Med. Genet. 8:378–380, 1971.
15. Finley, W. H., Finley, S. C. and Monsky, D.: An extra small metacentric chromosome in association with multiple congenital malformations. J. Med. Genet. 8:381–383, 1971.
16. Smith, K. D., Steinberger, E., Steinberger, A. and Perloff, W. H.: A familial centric chromosome fragment. Cytogenet. Cell Genet. 4:219–226, 1965.
17. Walzer, S., Breau, G. and Gerald, P. S.: A chromosome survey of 2400 normal newborn infants. J. Pediatr. 74:438–448, 1969.
18. Borgaonkar, D. S., Schimke, R. N. and Thomas, G. H.: Report of five unrelated patients with a small, metacentric, extra chromosome or fragment. J. Genet. Hum. 19: 207–222, 1971.
19. Tamburro, R. F. and Johnson, C. E.: An extra small metacentric chromosome in a female child. J. Med. Genet. 4:295–297, 1966.
20. Taft, P. D., Dodge, R. R. and Atkins, L.: Mental retardation and multiple congenital anomalies. Am. J. Dis. Child. 109:554–557, 1965.
21. Mukherjee, D. and Burdette, W. J.: A familial minute iso chromosome. Am. J. Hum. Genet. 18:62–69, 1966.
22. Hulten, M., Lindsten, J., Fraccaro, M. et al: Extra minute chromosome in somatic and germline cells of the same person. Lancet 2:22–24, 1966.
23. Pfeiffer, R. A., Diekmann, L. and Buchner, T.: Familial occurrence of a small metacentric extra chromosome with satellites on both ends. Ann. Genet. 10:124–130, 1967.
24. Hoehn, H., Reinwein, H. and Engal, W.: Genetic studies on a minute centric fragment transmitted through three generations. Cytogenet. Cell Genet. 9:186–198, 1970.
25. Abbo, G. and Zellweger, H.: The syndrome of the metacentric microchromosome. Helv. Paediatr. Acta 25:83–94, 1970.
26. Tangheroni, W., Cao, A. and Furbetta, M.: Multiple anomalies associated with an extra small metacentric chromosome: Modified Giemsa stain results. Humangenetik 18:291–295, 1973.
27. Crandall, B. F., Carrell, R. E. and Sparkes, R. S.: Chromosomes findings in 700 children referred to a psychiatric clinic. J. Pediatr. 80:62–68, 1972.

Ring Y Chromosome Without Mosaicism*

Miriam G. Wilson, MD , Ronald B. Stein, MD and Joseph W. Towner, PhD

Recently we had the opportunity to study a man whose karyotype showed a Y chromosome ring. This chromosome anomaly unassociated with mosaicism must be unusual since we know of only one other report in the literature, that of German and co-workers (1). We are reporting our patient to further elucidate the phenotype associated with a deleted Y chromosome and to present information about the location of male-determining genes on the Y chromosome.

CASE REPORT

A 53-year-old white man requested medical care because of hypogonadism and difficulty walking due to pain in his hips. He was previously seen in a hospital outpatient department for complaints of impotency and gynecomastia, where atypical Klinefelter syndrome was suggested and severe degenerative arthritis of the hips diagnosed. He was not aware that he had small testes and stated that potency was normal up to his mid-40s. He also stated that he shaved daily. He was first married one year ago and had not fathered any children, as far as he knew.

He gave conflicting accounts of his personal and medical history. He was the second born in a sibship of 9 children to young parents of northern European descent. Birthweight was 3175 gm at term. The patient was reared in Oklahoma up to 10 years of age, at which time the family moved to California. It appears that he was in the Marine Corps for about 30 years, probably as a private first-class, although at another time he said that he was a major in the Corps, which seemed unlikely in view of the general appearance of mild mental retardation. He was recently retired or honorably discharged.

*Supported in part by USPHS grant 286 from Maternal and Child Health Services.

Birth Defects: Original Article Series, Volume XII, Number 5, pages 105–112
© 1976 The National Foundation

Past medical history was not remarkable. There were no serious previous illnesses and no surgical procedures except for a tonsillectomy during childhood. He did not smoke, infrequently drank alcoholic beverages and took no medicines other than occasional aspirin. Specifically, he said that he had not received any testosterone injections.

The patient had 5 brothers, of whom 4 have normal children. One brother was married and had no children (reason unknown). Two of the brothers were approximately 6' tall and 3 were about 5'8". The proband was the shortest male in the sibship. The patient had 3 sisters, all with normal children. Family history of inherited disease, fetal deaths, congenital anomalies and mental retardation was denied.

The patient was a pleasant obese man who was cooperative and talkative; however, he seemed evasive and mildly mentally retarded. Height was 164 cm and weight was 92 kg. He walked slowly and painfully, using a cane. Blood pressure, pulse rate and respiratory rate were normal. His skin was soft and velvety to the touch. The facial skin showed fine wrinkles and a scanty beard. A moderate kyphosis and slight prominence of the right posterior hemithorax were noted. His breasts were large and pendulous (Fig. 1). Axillary and pubic hair was sparse. Testes were soft and small; the right measured 1.0 cm and the left 1.5 cm in the maximal diameters. The range of motion at both hips was limited and mild pitting edema of both legs was present.

Bilateral hip arthroplasties were done because of the incapacitating arthritis. Testosterone injections were given (100 mg of depo-testosterone every 2 weeks), after which the patient reported much improved potency and a heavier growth of beard. There was no change in his voice, which had previously been within the normal range.

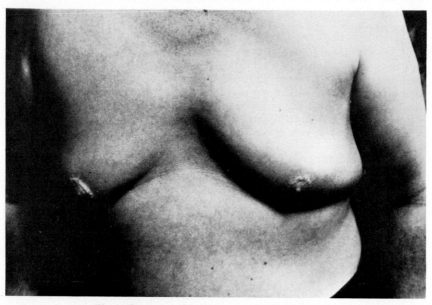

Fig. 1. Propositus. Marked gynecomastia was present.

Laboratory Data

Baseline plasma testosterone was decreased (145 ng% as compared to normal adult male values of 400–1000 ng%). Plasma FSH was 27.1 mIU/ml (normal adult male levels of plasma FSH 3–17 mIU/ml) and LH was 31.8 mIU/ml (normal adult male levels of plasma LH 5–22 mIU/ml). Normal results were obtained from routine blood chemistries (including calcium and phosphorus) and tests of thyroid function, liver function and peripheral blood counts. Alkaline phosphatase was elevated to 109 units. Serology was nonreactive. Radiograms demonstrated advanced degenerative changes in the hip joints and moderately involved vertebrae.

Chromosome analysis from peripheral blood cultures revealed a karyotype of 46,X,ring (Y). Although occasional random losses of other chromosomes were found, the Y chromosome was found in all 100 karyotypes examined. The ring Y was represented by a small mass of chromatin or sometimes paired dots which were no larger than the short arm of the normally expected Y and showed little variation in size in different karyotypes (Fig. 2). Although no hole in the center of the ring could be seen, the configurations of the chromosome were most consistent with ring formation. Preparations were examined by quinacrine-banding and Giemsa-banding. No bright fluorescence of the Y chromosome was observed, nor was it observed elsewhere in the karyotype (25 preparations examined). A Giemsa-banded preparation also demonstrated a normal karyotype other than the ring Y (Fig. 3). Specifically the X chromosome appeared normal. Examination of a buccal smear for Barr bodies gave negative results. Unfortunately it was not possible to obtain blood specimens for examination from the parents or a male sib of the propositus.

DISCUSSION

Polymorphism of the Y chromosome in normal males includes considerable variation in length ranging from shorter than one half the length of a G-group chromosome to as long as a D-group chromosome (2–5). It is generally agreed that the variable size of the Y chromosome in normal males represents variation in the length of the long arm, specifically in the length of the brightly fluorescing segment of the long arm. A very short Y chromosome without a prominent brightly fluorescing segment, therefore, cannot be detected by examination of interphase cells for fluorescing Y bodies.

We assume that the Y ring found in this patient represented only the centromere and a very small portion of chromatin from the proximal short and long arms, which united after deletions of both ends. In the instance reported here, no brightly fluorescing portion of the Y was observed on quinacrine-banding nor was it observed elsewhere in the karyotype. Since the karyotype of many normal males do not show the brightly fluorescing distal segment of the Y and no male relatives of the propositus were available for study, this observation in itself is insufficient to demonstrate a deletion. The ring Y, however, was very small, occasionally represented by a chromatin mass of less than 20% that of a No. 22 chromosome and comparable to the size of the short arm of the usually

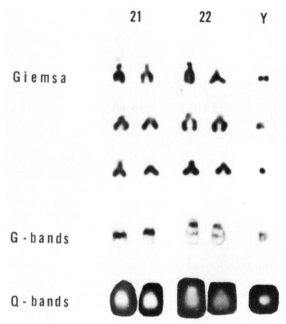

Fig. 2. Partial karyotypes from propositus demonstrating G-group chromosomes and the Y ring by standard Giemsa-staining, Giemsa-banding and quinacrine-banding.

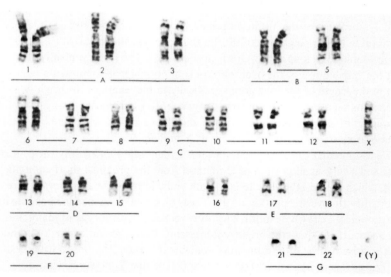

Fig. 3. Giemsa-banded karyotype from propositus demonstrating normal chromosomes except for the ring Y.

expected Y chromosome (Fig. 2, line 3). In spite of the dramatic decrease in size, the ring was stable and consistently represented in all cells examined.

Evidently a gene or genes for maleness and testicular differentiation were retained in this instance. Sufficient genetic material was present for testicular differentiation, yet insufficient for normal testicular development. The male phenotype of this patient, although sexually nonambiguous, was affected significantly by the loss of chromosome material; thus, it appears that some genes determining the male phenotype were not present or not expressed. Since this man was not short by general population standards (although short compared to parents and sibs), a gene or genes involved in stature possibly were retained within the small Y chromosome that was present. A relative deficiency in stature could be related to specific gene deletion, deficit of total sex chromosome material, decreased gonadal hormone or an undetected XO mosaicism.

The plasma testosterone was remarkably low, reflecting diminished hormone of testicular interstitial cell origin. Pituitary gonadotropins were high, as expected in primary hypogonadism. The testes were very soft — an observation which does not support the thesis (at least in this patient) that continued high gonadotropins result in hyalinized testicular tubules and fibrotic testes.

Structural abnormalities of the Y are rare occurrences. Since they are likely to be associated with mosaicism for a 45,X line, information regarding loci of genes on the Y chromosome is limited. There are 6 cases known to us of structurally abnormal Y chromosomes involving presumed deletions and without mosaicism for an XO line (1, 6–9). The first report of a deleted Y was by Vaharu et al (6), who described a small mentally retarded girl with an enlarged clitoris and a karyotype of 45,X plus an additional small piece of chromatin, presumably the Y chromosome. There is no information regarding the gonad histology nor the appearance of the father's Y. Jacobs and Ross (7) reported a presumptive isochromosome for the long arm of Y in each of 2 nonmasculinized females with amenorrhea and normal stature. The isochromosome Y was diagnosed on the basis of morphology in standard karyotypes, a negative X-chromatin result and a relatively late-labeling autoradiographic pattern consistent with DNA synthesis of a Y chromosome, but not so late as expected for a late-labeling X. In one of these instances, 2 male sibs of the patient were demonstrated to have a normal Y. No male relatives of the other patient were studied. Other cases of a metacentric Y chromosome, presumably an isochromosome for the long arm, were described in normal men (10, 11); consequently, the significance of a metacentric Y is uncertain, particularly in the early reports when no special staining was available. In 1965 Nakagome et al (8) described a severely retarded 2-year-old boy with normal male genitalia, whose karyotype showed 45 chromosomes and a minute fragment, presumably the Y chromosome, which was about the same size as the short arm of a G-group chromosome. No mosaicism was present; the 2 cells of the 29 cells examined without the marker Y chromosome were assumed to represent random losses. The father's Y chromosome was stated to be morphologically normal. No paternity tests were done. The ability to examine the karyotype of the propositus by quinacrine fluorescence was not present in 1965, and the possibility of translocation of Y chromatin to another chromosome could not be ruled out.

Most informative up to the present case regarding the localization of Y-borne genes are the reports by Armendares et al (9) of a nonmosaic dicentric Y chromosome and German et al (1) of a nonmosaic Y ring. Armendares and co-workers (9) described a patient with stigmata of Turner syndrome who had also epididymis tissue and gonadoblastoma in the gonadal streaks. The chromosome analysis from this patient showed a nonmosaic cell line of 46,X, dicentric Y, from which a portion of the short arm was apparently deleted on the basis of quinacrine-staining. Presumably male-determining genes, lost in this patient, were located on the deleted portion of the short arm and some were retained on the dicentric Y.

The phenotype of the patient of German et al (1) was similar to our patient and may profitably be compared in detail (Table 1). Both patients had micro-testes, gynecomastia, fine and smooth skin, sparse facial and body hair, and a short stocky build with apparently short limbs. The individual described by German et al (1) was an intelligent man with dwarfism, ie with a height less than the 3rd percentile. Our patient was probably mildly mentally retarded and, although short in comparison with his male sibs, had a height measurement between the 3rd and 10th percentiles for the general population. In both these instances, the ring Y was unusually small and represented in many cells by only a fragment of chromatin.

Ring Y chromosomes involving mosaicism with a 45,X line were described in 4 patients (12–15). It is noteworthy that testicular tissue was found in these cases. Apparently there were male determinants on the ring Ys which were expressed in the phenotype.

SUMMARY

A 53-year-old man with a short stocky build, mild mental retardation, gynecomastia and hypogonadism was found to have a small ring Y chromosome unassociated with mosaicism. The ring Y was represented by a minute portion of chromatin or sometimes paired dots which were no larger than the short arm of the normally expected Y. No brightly fluorescent segment of the ring Y was present nor was it observed elsewhere in the karyotype. A primary medical problem was severe osteoarthritis, necessitating bilateral hip arthroplasties. Plasma testosterone was markedly decreased and plasma gonadotropins were increased. Potency was improved following testosterone injections.

TABLE 1. Nonmosaic Patients With Ring Y

	Patient Reported by German et al (1)	Our Patient
Age (years)	49	53
Retardation	Not present	Probably mild
Voice	High-pitched	Not unusual
Stature	140 cm	164 cm
Body build	Wt 48.7 kg	Stocky, wt 92 kg
Skin	Soft, smooth	Soft, fine wrinkles
Hair on head	Full, no frontal recession	Not full, frontal recession
Eyebrows	Bushy	Not remarkable
Facial hair	Sparse	Sparse
Neck	Short	Short, broad
Axillary hair	Sparse	Sparse
Breasts	Palpebral breast tissue	Breast enlargement
External genitalia	Short penis (5 cm), normally formed	Penis and scrotum not remarkable
	Pubic hair luxuriant, female distribution	Pubic hair of female distribution
	Testes in scrotum, 1.5–2.0 cm diameter, normal consistency	Testes in scrotum, 1.0–1.5 cm diameter, soft
Limbs	Appear short. Bowed legs	Appear short. Severe osteoarthritis
Laboratory:		
Gonadotropins	Normal	Elevated
Testosterone (plasma)	–	Decreased
Thyroid function	Normal	Normal

We conclude that genes responsible for testicular differentiation, maleness and possibly height are located close to the centromere of the Y chromosome, possibly on both the short and long arm. We also conclude that multiple genes are required for a fully developed male phenotype and apparently some of these genes were deleted or not expressed in this patient.

REFERENCES

1. German, J., Simpson, J. L. and McLemore, G. A.: Abnormalities of human sex chromosomes. I. A ring Y without mosaicism. Ann. Génét. 16:225–231, 1973.
2. Court Brown, W. M.: "Human Population Cytogenetics." New York: John Wiley & Sons, 1967, pp. 56–57.
3. Borgaonkar, D. S., McKusick, V. A., Herr, H. M. et al: Constancy of the length of the human Y chromosome. Ann. Génét. 12:262–264, 1969.
4. Lubs, H. A. and Ruddle, F. H.: Applications of quantitative karyotyping of chromosome variation in 4400 consecutive newborns. In Jacobs, P. A., Price, W. H. and Law,

P. (eds.): "Human Population Cytogenetics," Baltimore: Williams & Wilkins, 1970, pp. 119–142.

5. Meisner, L. F. and Inhorn, S. L.: Normal male development with Y chromosome long arm deletion (Yq–). J. Med. Genet. 9:373–377, 1972.

6. Vaharu, T., Patton, R. G., Voorhees, M. L. and Gardner, L. I.: Gonadal dysplasia and enlarged phallus in a girl with 45 chromosomes plus "fragment." Lancet 1:1351, 1961.

7. Jacobs, P. A. and Ross, A.: Structural abnormalities of the Y chromosome in man. Nature 210:352–354, 1966.

8. Nakagome, Y., Sasaki, M., Matsui, I. et al: A mentally retarded boy with a minute Y chromosome. J. Pediatr. 67:1163–1167, 1965.

9. Armendares, A., Buentello, L., Salamanca, F. and Cantu-Garza, J. M.: A dicentric Y chromosome without evidence of sex chromosomal mosaicism, 46,XYqdic, in a patient with features of Turner's syndrome. J. Med. Genet. 9:96–131, 1972.

10. Solomon, I. L., Hamm, C. W. and Green, O. C.: Isochromosome Y in a father and his mentally retarded son who had sex chromosome mosaicism. N. Eng. J. Med. 271:586–592, 1964.

11. Stoeckenius, M.: A probable Y isochromosome in father and son of tall stature. Lancet 2:391, 1966.

12. Sarto, G. E., Opitz, J. M. and Inhorn, S. L.: Considerations of sex chromosome abnormalities in man. In Benirschke, K. (ed.): "Comparative Mammalian Cytogenetics." Berlin: Springer-Verlag, 1969, pp. 390–413.

13. Chandley, A. C. and Edmond, P.: Meiotic studies on a subfertile patient with a ring Y chromosome. Cytogenet. Cell Genet. 10:295–304, 1971.

14. Khudr, G. and Benirschke, K.: Y ring chromosome associated with gonadoblastoma in situ. Obstet. Gynecol. 41:897–901, 1973.

15. de Almeida, J. C. C., Barcinski, M. A., do Céu Abreu, Maria et al: 45,X/46,Xr(Y) in a case of asymmetrical testicular differentiation. Ann. Génét. 17:37–40, 1974.

Tertiary Trisomy 14: Is There a Syndrome?

Sergio D. J. Pena, MD, M. Ray, PhD, P. J. McAlpine, PhD,
C. Ducasse, MD, FRCP(C), J. Briggs, MD, FRCP(C)
and John L. Hamerton, DSc

Trisomy 14 is not an uncommon chromosomal aberration. It is present in at least 2.5% of all spontaneous abortions (1) and consequently, assuming that 15% of pregnancies terminate in miscarriages (2), a minimum of 0.4% of all conceptuses are trisomic for chromosome 14. This trisomic state is apparently highly lethal during the prenatal period, with very few of these conceptuses reaching term. One postnatal case of trisomy 14 has been reported (3); in this case identification of the extra autosome was done by autoradiography. In contrast, duplication of segments of chromosome No. 14 due to chromosome rearrangements seems to be more common in postnatal life (4–12).

We report here 2 sisters displaying a similar pattern of congenital anomalies and presenting with an identical chromosome rearrangement resulting in duplication of a segment of chromosome 14 and a deletion of a segment of chromosome 13, a [46, XX, − 13, der (14),+t (13; 14) (q13;q22) mat] karyotype. Their phenotype will be compared to that of the other reported cases.

CASE REPORTS

Case 1

The proband is an Amerindian girl who was admitted to the Children's Centre, Winnipeg, at the age of 20 months. She was referred from the northern Manitoba community of Wabowden for investigation of multiple congenital anomalies and recurrent lower respiratory infections. Her birthweight was 2600 gm and the pregnancy and her delivery were uneventful. On admission her height was 65 cm, weight 5800 gm and head circumference 42 cm—

Birth Defects: Original Article Series, Volume XII, Number 5, pages 113–118
© 1976 The National Foundation

all of which were below the 3rd percentile for her age. Physical examination revealed an irritable, slightly hypertonic infant with a flat occiput, low-set malformed ears, mild blepharophimosis, a posterior median cleft of the palate and
a small chin (Fig. 1). There was camptodactyly and bilateral overlapping of the
hallux by the second toe. A large umbilical hernia and a presacral dimple were
also present. The child died at 29 months of severe pneumonia following an
aspiration episode. Necropsy showed a normal heart; neuropathology was not
performed.

Case 2

The sister of the proband was born at term with a weight of 1800 gm after
a normal pregnancy and delivery. At age 3½ months her height was 47 cm, weight
was 2600 gm and head circumference 34 cm — all below the 3rd percentile for her
age. Physical examination showed low-set but otherwise normal ears, severe
blepharophimosis, a median cleft in the soft palate and micrognathia (Fig. 2).
There was mild camptodactyly , an umbilical hernia and a presacral dimple.
Repeated lower respiratory infections led to death at 5½ months. A summary
postmortem examination revealed a normal heart.

The parents are North American Indians and apparently unrelated. The
mother is healthy and was 26 and 27 years old, respectively, at birth of the 2
infants. There was no family history of birth defects.

CHROMOSOME STUDIES

Chromosome analysis on peripheral blood lymphocytes and skin fibroblasts
of the proband and her sister using conventional staining, Q- and G-banding revealed a 46,XX,−13,der(14),t(13;14)(q12;q22) mat karyotype. Analysis of
the mother's chromosomes showed her to be carrying a balanced reciprocal

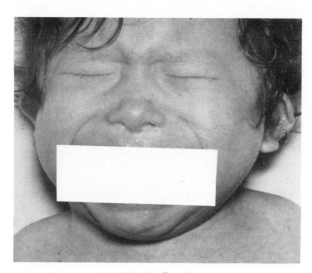

Fig. 1. *Case 1.*

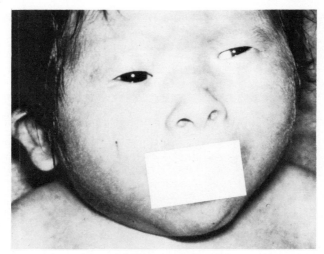

Fig. 2. *Case 2.*

translocation between chromosome Nos. 13 and 14 [46,XX,t(13;14)(q12;q22)] (Fig. 3), while the father had normal chromosomes (46,XY). Chromosome studies on other family members have so far not proved possible.

From this analysis it is clear that both infants had inherited the der(14) chromosome from their mother and that each had a deficiency of the segment (13)(pter→q12) and a duplication of the segment (14)(pter→q22).

GENETIC MARKER STUDIES

In an attempt to look for apparent hemizygous expression of gene loci due to the deficiency of a segment of one homologue of chromosome 13, genetic marker studies were carried out on hemolysates and white blood cell lysates of the parents and on fibroblast lysates of *Case 2* to determine enzyme phenotypes related to 22 gene loci. Fibroblasts derived from *Case 1* were not available for genetic marker studies. The phenotype of each of the markers examined in *Case 2* was consistent with the phenotypes that could be expected on the basis of the parental phenotypes. There was no evidence of hemizygosity at any of the gene loci tested that have not been given a chromosome assignment or that have been assigned to chromosome 13 in man. No evidence to support or refute the tentative assignment of the esterase D (*Es D*) gene locus to chromosome 13 in man (13) was obtained as the mating was uninformative for studying the segregation of alleles at the *Es D* locus.

DISCUSSION

The reciprocal translocation carried by the mother is unequal and involves 2 acrocentric chromosomes and should result in an asymmetric pachytene configuration with short interstitial segments and, hence, a low probability of the

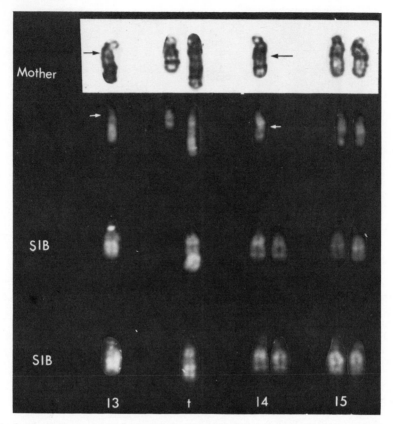

Fig. 3. Partial karyotype showing balanced translocation in mother [t(13;14)(q12;q22)] and der(14) in her 2 sibs. The chromosome identified by t in the figure is the der(14).

occurrence of an interstitial crossover. It has been suggested that these conditions might lead to a high frequency of adjacent homologous segregation (14). This is confirmed in the present instance by the finding of 3 chromosome 14 centromeres in the karyotype of the proband and her sister.

In the 2 infants reported here, the segment (pter→q22) of chromosome 14 was duplicated. A similar duplication was observed in 7 other reported cases (4, 5, 8, 10–12), while 2 other subjects carried a smaller duplication (pter→ q12)(6, 7). In contrast, the case described by Pfeiffer et al (9) had a duplication of the terminal part of the long arm (q24→qter) and had a very different phenotype. This case will not be discussed further.

Phenotypic comparison of the above cases, as well as of the patient with trisomy 14 (3), allows the clear definition of a clinical syndrome consisting of feeding difficulties, recurrent respiratory infections, failure to thrive associated with psychomotor retardation and a complex spectrum of malformations including low-set ears (8/12), microcephaly (8/12), broad flat nasal bridge (8/12), blepharophimosis and/or microphthalmia (7/12), micrognathia, (7/12), camptodactyly (7/12), cryptorchidism (2/14), clefts of palate or alveolar ridge (4/12), long philtrum (4/12), hypotelorism (3/12), seizures (3/12), rib anomalies (3/12) and clubfeet (3/12) (Fig. 4). Significant phenotypic variability exists, the most common malformations being present in more than 67% of affected children. This variability may be partly attributed to differences in the size of the duplicated segment and perhaps as well to the associated small deletions found in other autosomes. Nevertheless, the fact that, in 2 separate instances [Surana et al (11) and the present report], related patients displaying identical chromosomal abnormalities differed in their clinical features is evidence of the high degree of phenotypic variability in this disorder. To a greater or lesser extent, such variability is characteristic of most chromosome abnormalities. It is also relevant to note that the pattern of malformation lacks any distinctive feature — indeed, most of the malformations in the syndrome are rather nonspecific, being also components of a number of other conditions.

In conclusion, a partial trisomy 14 syndrome does exist. However, the observed phenotypic variability and the lack of any pathognomonic or distinctive abnormalities leads us to believe that definition of this syndrome will be of only limited value for the clinical recognition of affected individuals.

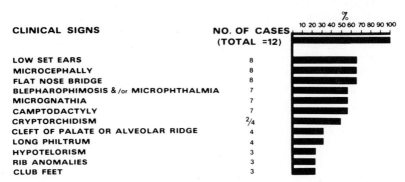

Fig. 4. Diagram depicting the more important dysmorphic features of the partial trisomy 14 syndrome.

REFERENCES

1. Kajii, T., Omaha, K., Niikawa, N. et al: Banding analysis of abnormal karyotypes in spontaneous abortion. Am. J. Hum. Genet. 25:539, 1973.
2. Boué, A. and Boué, I.: Évaluation des erreurs chromosomiques au moment de la conception. Biomedicine 18:372, 1973.

3. Murken, I. D., Bauchinger, M., Palitzsch, D. et al: Trisomie D_2 bei einem 2½ Jährigen Mädchen (47,XX,14 +). Humangenetik 10:254, 1970.
4. Allderdice, P. W., Miller, O. J., Miller, O. A. et al: Familial translocation involving chromosomes 6, 14 and 20 identified by quinacrine fluorescence. Humangenetik 13:205, 1971.
5. Fawcett, W. A., McCord, W. K., and Francke, U.: Trisomy 14q −. In Bergsma, D. (ed.): "New Chromosomal and Malformation Syndromes," Birth Defects: Orig. Art. Ser., vol. XI, no. 5. Miami: Symposia Specialists for The National Foundation − March of Dimes, 1975, pp. 223−228.
6. Fryns, J. P., Cassiman, J. J. and van den Berghe, H.: Tertiary partial trisomy 47,XX, +14q −. Humangenetik 24:71, 1974.
7. Laurent, C., Dutrillaux, B., Biemont, M.-Cl. et al: Translocation t (14q −; 21q +) chez le peré. Trisomie 14 et monosomie 21 partielles chez la fille. Ann. Genet. (Paris) 16: 281, 1973.
8. Muldal, S., Enoch, B. A., Ahmed, A. and Harris, R.: Partial trisomy 14 and pseudo-xanthoma elasticum. Clin. Genet. 4:480, 1973.
9. Pfeiffer, R. A., Büttinghaus, K. and Struck, H.: A balanced reciprocal translocation t (14q −; 21q +). Humangenetik 20:187, 1973.
10. Short, E. M., Solitaire, G. B. and Breg, W. R.: A case of partial 14 trisomy 47,XY, (14q −)+ and translocation t (9p +;14q −) in mother and brother. J. Med. Genet. 9: 367, 1972.
11. Surana, R. B., Conen, P. E., Braudo, M. and Keith, J. D.: Familial tertiary trisomy with +(14q −;1 +?). In Bergsma, D. (ed.): "Clinical Cytogenetics and Genetics," Birth Defects: Orig. Art. Ser., vol. X, no. 8. Miami: Symposia Specialists for The National Foundation − March of Dimes, 1974, pp. 53−65.
12. Reiss, J. A., Wyandt, H. E., Magenis, R. E. et al: Mosaicism with translocation: Auto-radiographic and fluorescent studies of an inherited reciprocal translocation t (29 +;14q −). J. Med. Genet. 9:280, 1972.
13. Van Heyningen, V., Bobrow, M., Bodmer, W. F. et al: Chromosome assignment of some human enzyme loci: Mitochondrial malate dehydrogenase to 7, mannosephosphate isomerase and pepruvate kinase to 15 and probably, esterase D to 13. Ann. Hum. Genet. 38:295−303, 1975.
14. Hamerton, J. L.: "Human Cytogenetics." New York: Academic Press, 1971, vol. 1.

Partial Trisomy of Chromosome 14: (+14q−)*

Gentry W. Yeatman, Maj, MC, USA **and Vincent M. Riccardi,** MD

Numerous patients with an extra, small acrocentric (ESA) chromosome but without the characteristic features of Down or Klinefelter syndromes have been described. These chromosomes have been said to represent a No. 22 or a portion of various other autosomes. With the advent of chromosomal banding techniques many ESAs have been shown to originate from D-group chromosomes, resulting in partial trisomies 13, 14 and 15 (1−8). The following case is presented to emphasize the paucity and nonspecificity of clinical findings associated with a 47,+14q− karyotype and underscore the need for banding techniques to be applied routinely. It is of interest that increased fragility of the proximal long arm of chromosome 14 has been demonstrated in vitro (9).

CASE REPORT

The proband was the product of a term pregnancy complicated by threatened abortion, edema, hypertension, gastroenteritis and allergic rhinitis. During her pregnancy the mother received Enovid-E (Searle; norethynodrel 2.5 mg and mestranol 0.1 mg) meprobamate, chlorthiazide, chlorpheniramine maleate and Co-Pyronil (Lilly; pyrrobutamine compound). Total weight gain was 7700 gm. A decrease in fetal motion was noted during the last trimester.

The proband has 3 sibs. They are all of normal intelligence and height but have mild scoliosis. Polycystic kidney disease is suspected in 1 sib and 2 maternal relatives died of purported polycystic kidney disease. The mother's sibship included 1 stillborn and 1 child who died shortly after birth with cardiorespiratory anomalies. There is no history of stillbirths, abortions or severe anomalies in the proband's sibship. A paternal great-aunt had twin sisters who were said to look just like the proband; they were retarded and died before age 11 years.

*Supported in part by the Colorado–Wyoming Regional Medical Program, The National Foundation – March of Dimes and the Henry J. Kaiser Family Foundation.

The opinions or ascertainments contained herein are the private views of the authors and are not to be construed as official, or as representative of the views of the Department of the Army or the Department of Defense.

Birth Defects: Original Article Series, Volume XII, Number 5, pages 119–124

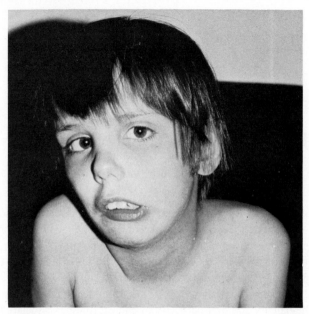

Fig. 1. Proband. The face is long and thin; the mouth is held open and a distracted, relatively expressionless appearance is noted. However, she is capable of full facial expression, and she is strikingly similar in appearance to other family members.

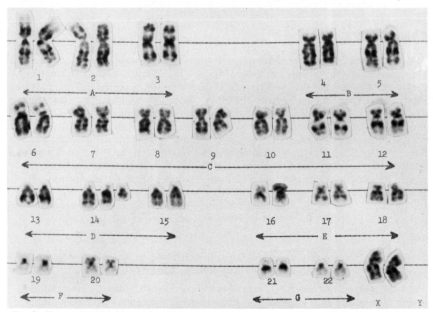

Fig. 2. Karyotype of the proband. All cells counted showed 47,XX,+14q− chromosome constitution. Breakpoint was at 14q22.

Birthweight was 2700 gm and length was 48 cm. She was a floppy infant with a weak cry, poor suck and feeding difficulties. At 3 months she was not developing as rapidly as previous sibs. At 5 months a clockwise rotary nystagmus was noted; an EEG was nonconclusive. At 10 months she was given the diagnosis of cerebral palsy; a repeat EEG was normal. She first smiled at 5 months, held a bottle at 11 months, sat with assistance at 13 months, crawled at 19 months, pulled to stand at 21 months, furniture-walked at 33 months and developed a one word vocabulary at about 4 years. Peripheral blood karyotypes at ages 3 and 7 showed an ESA chromosome. Due to lack of physical stigmata of Down syndrome, the diagnosis of trisomy 22 was given.

At 8 years, height, weight and head circumference were below the 3rd percentile. She showed a small carp-like mouth, moderate overbite, bilateral ptosis, hyperextensible joints, hypotonia and nystagmus. An EEG showed disorganized right central spikes. An IVP was normal.

At 11 and 12 years several additional findings were noted (Fig. 1): prominent nasal bridge, hyperopia, alternating esotropia, periorbital hyperpigmentation, micrognathia, horizontal ridging of the teeth, prominent lips, narrow shoulders and asymmetry of the chest. The ocular fundi, irides, auricles and limbs were normal. Dermatoglyphics were normal, all fingertips showing whorls or ulnar loops. The hallucal regions showed loop distal patterns. Patellas were present. There was hypotonia with a generalized decrease in muscle mass. The breasts and external genitalia were prepubescent.

Trypsin G-banding of peripheral lymphocytes showed 47 chromosomes in all of 20 cells counted. The ESA chromosome resembled the short arm, centromere and proximal long arm of chromosome 14 (Fig. 2). The karyotype was designated 47,XX, +14q − with the breakpoint at q22. Karyotypes of the father, mother and 2 sibs were normal. A previous karyotype of the remaining sib had been reported as normal.

DISCUSSION

There are 2 general areas of importance in considering the present case: clinical and cytogenetic. First, she shows virtually no distinctive findings; in comparison both with patients having an unidentified ESA (10–13) and those with a + 14q − karyotype (1–6), the clinical features are quite nonspecific. Second, the cytogenetic analyses of patients with an ESA are just beginning to sort out the confusion which developed in the prebanding era. As the origin of each ESA is being established, different chromosomes are being implicated (1–8) and certain breakpoints are being shown to be nonrandom in their occurrence. This is true for chromosome 14: in vivo at 14q22 (eg Pfeiffer et al (14) and present case) and in vitro at 14q12 (9). In addition, although ESA cases are relatively rare, there are a number of reports associating an ESA with another separate diagnosis (3, 12, 15–17) and thus the pathogenetic import of the extra chromosomal material is not always clear.

Our patient strikingly resembles her parents and sibs, indicating that she does not have a physiognomy and facies solely or mainly determined by chromosome constitution. The ptosis is mild, and this and other findings (small mouth, nystagmus, hypotonia, etc) are nonspecific (Table 1). Her purported resemblance

TABLE 1. Comparison of Features From the Present Case and Previously Reported Cases of +14q−

	Present Case	Short et al (1)	Reiss et al (2)	Muldal et al (3)	Laurent et al (4)	Fryns et al (5)	Allderdice et al (6)*
Maternal age	36	22	21	31	29	18	27
Paternal age	?	27	26	31	32	20	27
Sex	F	M	M	F	F	F	F
Gestational age	term	term	term	0	term	term	term
Birthweight	2700 gm	2910 gm	2400 gm	0	2100 gm	2200 gm	2600 gm
Mental retardation	+	0†	+	+	0	+	+
Growth retardation	+	0†	+	+	+	+	+
Microcephaly	+	+	+	+	+	+	+
Micrognathia	+	+	+	0	−	0	+
Ear malposition	−	+	+	+	−	0	+
Abnormal cranium	−	+	0	0	+	0	0
Seizures/abnormal EEG	+	+	0	+	−	0	0
Hypotonia	+	+	−	0	−	−	0
Cardiovascular anomaly	−	+	+	+	−	0	0
High-arched palate	+	−	0	+	0	0	−
Frequent infections	+	0†	0	+	0	0	+
Abnormal cry	?	+	+	0	0	0	+
Finger contractures	−	+	+	0	+	0	+
Clinodactyly	+	0	0	+	−	0	0

+ = Present
− = Absent
0 = Insufficient data
*Possibly a portion of 20p also present: 47,XX,−6,+t(6q;20p?),+t(14q;16q)
†Neonatal death

to retarded distant relatives probably reflects her family resemblance more than anything. This consideration, in the context of the general language used to describe the abnormalities found in other cases with similar chromosome findings (Table 1), further emphasizes the nonspecificity of the features noted in our patient. There were virtually no features which would have ordinarily suggested a chromosomal etiology.

In 6 previous cases where the ESA has been identified as 14q− (1−6), the findings have included the following (Table 1): mental retardation, growth retardation, microcephaly, generalized hypotonia, feeding difficulties, abnormal cry, micrognathia, cardiac or vascular anomalies, ear malposition, high-arched palate, abnormal cranial shape, seizures and/or abnormal EEG and frequent infections. Bony anomalies are common with finger contractures, absent radius, abnormal ribs and clinodactyly. Facial findings include a wide nasal bridge, deep-set eyes, prominent lips, a held-open mouth and an expressionless face. There may be minimal physical findings, with growth and mental retardation being most obvious. The overlap and nonspecificity of such findings is further suggested by the review of Gustavson et al (10). The only consistently recognizable ESA disorder has been the cat-eye syndrome (13): iris colobomata, imperforate anus, preauricular sinuses or tags, and urinary tract anomalies. However, the involved chromosome has not been identified and no patients with partial trisomy 14 have been described with these findings.

The extra chromosomal material has been derived from a parental balanced reciprocal translocation in 5 of the 6 previously reported patients with a +14q− karyotype (1, 2, 4−6). In our patient there was no familial translocation. In only one case was mosaicism documented (2). Complete trisomy 14 is unknown except in spontaneous abortions (18).

SUMMARY

Partial trisomy 14 as a 47,+14q− karyotype is compatible with life and this state is distinguished on cytogenetic, not clinical, grounds. Clinical features are nonspecific and these children are most obvious because of growth and mental retardation.

This extends and broadens the indications for chromosome analysis. The significance of finding a +14q− karyotype must, however, be considered in the knowledge that ESA fragments have been associated with disorders that do not have a chromosome aberration pathogenesis. A balanced translocation should always be sought in parents of affected children. Finally, the proximal long arm of chromosome 14 apparently has at least 2 sites of relative fragility: 14q12 in vitro and 14q22 in vivo. Blood group analyses have been uninformative (2, 5).

REFERENCES

1. Short, E. M., Solitaire, G. B. and Breg, W. R.: A case of partial 14 trisomy 47,XY, (14q −)+ and translocation t(9p+;14q −) in mother and brother. J. Med. Genet. 9: 367–373, 1972.
2. Reiss, J. A., Wyandt, H. E., Magenis, R. E. et al: Mosaicism with translocation: Autoradiographic and fluorescent studies of an inherited reciprocal translocation t(2q+; 14q−). J. Med. Genet. 9:280–286, 1972.
3. Muldal, S., Enoch, B. A., Ahmed, A. and Harris, R.: Partial trisomy 14q − and pseudoxanthoma elasticum. Clin. Genet. 4:480–489, 1973.
4. Laurent, C., Dutrillaux, B., Biemont, M. C. et al: Translocation t (14q−;21q+) chez le pere. Trisomie 14 et monosomie 21 partielles chez la fille. Ann. Genet. (Paris) 16: 281–284, 1973.
5. Fryns, J. P., Cassiman, J. J. and Van den Berghe, H.: Tertiary partial 14 trisomy − 47,XX,+14q −. Humangenetik 24:71–77, 1974.
6. Allderdice, P., Miller, O. J., Miller, D. A. et al: Familial translocation involving chromosomes 6, 14 and 20 identified by quinacrine fluorescence. Humangenetik 13:205–209, 1971.
7. Escobar, J. L. and Yunis, J. J.: Trisomy for the proximal segment of the long-arm of chromosome 13. A new entity? Am. J. Dis. Child. 128:221–222, 1974.
8. Watson, E. J. and Gordon, R. R.: A case of partial trisomy 15. J. Med. Genet. 11:400–402, 1974.
9. Hecht, F., McCaw, B. K., Peakman, D. and Robinson, A.: Non-random occurrence of 7–14 translocations in human lymphocyte cultures. Nature 255:243–244, 1975.
10. Gustavson, K. H., Hitrec, V. and Santesson, B.: Three non-mongoloid patients of similar phenotype with an extra G-like chromosome. Clin. Genet. 3:135–146, 1972.
11. Zellweger, H., Mikamo, K. and Abbo, G.: Two cases of multiple malformations with an autosomal chromosome aberration − Partial trisomy D? Helv. Paediatr. Acta 17:290–295, 1962.
12. Giorgi, P. L., Paci, A. and Ceccarelli, M.: An extra chromosome in a case of Tay-Sachs disease with additional abnormalities. Helv. Paediatr. Acta 22:28–35, 1967.
13. Fryns, J. P., Eggermont, E., Verresers, H. and Van den Berghe, H.: A newborn with the cat-eye syndrome. Humangenetik 15:247–248, 1972.
14. Pfeiffer, R. A., Buttinghaus, K. and Struck, H.: Partial trisomy 14 following a balanced reciprocal translocation t(14q−;21q+). Humangenetik 20:187–189, 1973.
15. Becak, W., Becak, M. L., Andrade, J. D. and Manissadjian, A.: Extra acrocentric chromosome in a case of giant cavernous hemangioma with secondary thrombocytopenia. Lancet 2:468, 1963.
16. Hayward, J. D. and Bower, B. D.: Chromosomal trisomy associated with the Sturge-Weber syndrome. Lancet 2:844–846, 1960.
17. Fitzgerald, P. H.: A possible supernumerary chromosome associated with dystrophia myontonica. Lancet 2:456, 1962.
18. Lucas, M., Wallace, I. and Hirschhorn, K.: Recurrent abortions and chromosome abnormalities. J. Obstet. Gynaecol. Br. Commonw. 79:1119, 1972.

Deletion of 11q: Report of Two Cases and a Review*

Steven A. Larson, Cpt, MC, USA, Gentry W. Yeatman, Maj, MC, USA
and Vincent M. Riccardi, MD

As an increasing number of chromosome disorders are delineated (1), 2 important clinical observations emerge: affected persons may show only a paucity of physical findings and what findings there are may be nonspecific. Two patients with long-arm deletions of a chromosome 11 are presented to emphasize these points and thereby broaden the general indications for chromosome analysis by banding techniques. Three other patients with 11q– karyotypes have been reported previously (2–4). The most constant features of these patients have been mental and growth retardation, telecanthus and congenital heart disease. More variable features have included microcephaly, scaphocephaly, ptosis, epicanthal folds, abnormal slant to the palpebral fissures, micrognathia, carp mouth, simian creases and brachydactyly.

CASE REPORTS

Case 1

This white female was first seen at 7 months of age for evaluation of developmental and growth retardation and an unusual appearance. She was born at term to a 34-year-old gravida 6 mother. The father was age 39. All 5 previous pregnancies had produced normal children. The pregnancy was unremarkable, without illnesses or teratogenic exposures, and the delivery was uneventful, with a birthweight of 3480 gm. There was no family history of mental retardation or congenital

*Supported in part by the Colorado-Wyoming Regional Medical Program, The National Foundation – March of Dimes and the Henry J. Kaiser Family Foundation.

The opinions or ascertainments contained herein are the private views of the authors and are not to be construed as official, or as representative of the views of the Department of the Army or the Department of Defense.

Birth Defects: Original Article Series, Volume XII, Number 5, pages 125–130
© 1976 The National Foundation

malformations. By 7 months of age it was obvious that there was gross motor delay and failure to grow, and at that time she was functioning at a 1—2 month level. Her history was otherwise unremarkable except for frequent upper respiratory infections. On physical examination she was small for her age and hypotonic. She had a round face with right ptosis, telecanthus and a carp mouth. The hands and fingers were short, more marked on the right. The left foot was 2 cm shorter than the right; there was no other obvious asymmetry. The external genitalia were normal. There were no simian creases nor abnormal palmar or hallucal dermatoglyphic patterns. Fingertip patterns were as follows: whorl patterns for right 2 and 3 and left 2; all others showed ulnar loops. The heart was normal by auscultation, chest radiograph and ECG. Routine laboratory parameters were normal, as were thyroid function and urinary excretion of organic and amino acids and mucopolysaccharides. Radiologic studies including IVP were normal except for a delayed bone age of 3 months and comparative shortening of the metacarpals of the right hand and metatarsals of the left foot. An EEG revealed low voltage spindles on the left and almost absent spindle formation on the right. A nonbanded karyotype was interpreted as normal.

At 14 months the same findings were noted, with continued profound muscular hypotonia. Developmentally she functioned at the 4—5 month level at that time and her bone age was 9 months. Peripheral leukocytes were studied with trypsin G-banding and revealed a 46,XX,del(11) (q23) karyotype (Fig. 1). At 23 months (Fig. 2) her physical exam was unchanged except that her ptosis was now

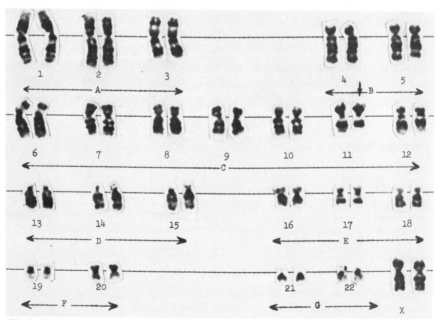

Fig. 1. Karyotype of *Case 1:* 46,XX,del(11) (q23). Note loss of the terminal lightly stained region of one of the No. 11 chromosomes (arrow). Exactly the same findings were noted for *Case 2.*

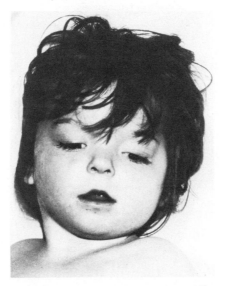

Fig. 2. *Case 1.* Note the bilateral ptosis, telecanthus, long philtrum and carp mouth.

bilateral and the philtrum was noted to be long. Her weight was 9.6 kg (3rd%),
height 83 cm (10th%) and occipitofrontal head circumference 46 cm (< 3rd%).
Developmental function was at the 6 month level with gross motor abilities
of 4 months. Her bone age was delayed at 15 months. It was also noted that
she was inattentive to almost all visual stimuli, although her pupils reacted directly
and consensually to light. Funduscopic exam was normal.

Chromosome analysis on the mother revealed a 46,XX karyotype by trypsin
G-banding. Paternal karyotypes have not yet been done.

Case 2

This patient, a 15-year-old female, was born of Spanish-American parents
when the mother was age 15 and the father 19. Pregnancy was unremarkable for
illnesses or teratogenic exposures. Delivery was at an estimated 26–28 weeks'
gestation in an automobile on the way to the hospital. The infant was said to
have been blue for 5–10 minutes before reaching the hospital. Birthweight was
1380 gm. The only abnormality noted at birth was a systolic heart murmur, later
shown to be due to a VSD. Major motor seizures developed at 3 days of age. By
6 months of age severe retardation and cortical blindness were documented; no
dysmorphic features were noted. Upon admission to an institution at age 4 she
was described as a rather pretty little girl. Prior to that time there had been
multiple episodes of acute infection, primarily repeated otitis media and
pneumonia. A chromosome analysis by conventional methods in 1965 (5 years
old) was interpreted as normal. At age 15 years (Fig. 3) she was severely retarded,
with an IQ estimated to be less than 10. Her weight at that time was 17.5 kg

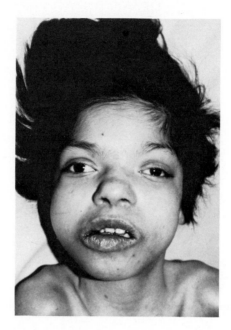

Fig. 3. *Case 2.* Note the telecanthus, hirsute forehead and prominent lower lip.

(50th% for a 4-year-old) and her head circumference was 42 cm (50th% for a 5-year-old). She had severe flexion contractures of all major limb joints secondary to spastic quadriplegia and she still experienced occasional major motor seizures. There were no findings, except telecanthus, to suggest a specific syndrome or chromosome disorder. Her forehead was hirsute and her lips were prominent, especially the lower (both findings possibly secondary to chronic Dilantin therapy), and there was an intermittent right exotropia. The breasts and external genitalia were prepubertal but otherwise normal. There were no simian creases nor abnormal palmar dermatoglyphics. Whorl patterns were present on fingers 2–4 of the right hand and finger 4 of the left; the remainder showed ulnar loops. Both hallucal regions showed loop distal patterns. Routine laboratory data and an IVP were unremarkable. Urinary amino acid screening was normal. A repeat chromosome analysis of peripheral lymphocytes with trypsin G-banding revealed a 46,XX,del(11)(q23) karyotype (Fig. 1). Permission for family studies was denied. The family history was said to be negative.

DISCUSSION

The major pupose of presenting these 2 patients is to emphasize the paucity and/or nonspecificity of findings beyond mental retardation which may be associated with a chromosomal aberration. Coincidentally, the improved resolution of chromosome banding techniques is demonstrated: both patients had been said to be chromosomally normal by previous nonbanding techniques [also see Linarelli et al (4)]. In *Case 1* the facial dysmorphism prompted a search for a syndrome diagnosis but the particular features were neither specific or striking nor associated with other anomalies except the smaller right hand and left foot. In *Case 2* there was virtually nothing to suggest a chromosome disorder and cerebral anoxia at birth provided a good explanation of her retardation. In retrospect, however, her telecanthus combined with a VSD seem noteworthy.

With our 2 patients, there are now 5 patients reported to have a terminal deletion of chromosome 11. There is no apparent association of advanced parental age (Table 1). Four of the patients were female, and 2 were premature. Mental and growth retardation were common to all 5 patients; telecanthus was noted in 4 and probably was present in the fifth as observed from the published photo (3). Three patients had a VSD, with or without other defects, and a fourth was

TABLE 1. Comparison of Features of Patients With the 11q– Syndrome

	Case 1	Case 2	Jacobsen et al (2)	Faust et al (3)	Linarelli et al (4)
Sex	Female	Female	Female	Female	Male
Age	23 mo	15 yr	14 mo (died)	9 yr	12 yr
Maternal Age	34	15	24	30	32
Paternal Age	38	19	25	30	33
Gestation	Term	26–28 wk	35 wk	Term	38 wk
Birthweight	3480 gm	1830 gm	2000 gm	?	2353 gm
Mental Retardation	+	+	+	+	+
Growth Retardation	+	+	+	+	+
Telecanthus	+	+	+	?	+
Abnormal Palpebral Slant	+ ($\downarrow$)	–	+ ($\uparrow$)	0	+ ($\downarrow$)
Ptosis	+	–	0	0	+
Carp Mouth	+	–	0	–	+
VSD	–	+	+*	?†	+
Brachydactyly	+	–	+	0	+
Microcephaly	+	+	+	0	0
Scaphocephaly	–	–	+	0	0
Micrognathia	–	–	+	0	+
Epicanthic Folds	–	–	+	–	+
Simian Creases	–	–	+	0	+

Note the similarities are either nonspecific (eg mental retardation, telecanthus, VSD) or limited and subtle.

+ = Present
– = Absent
? = Possibly present

0 = Insufficient data
*VSD, ASD and pulmonary hypertension
†Murmur only

reported to have a murmur. An abnormal slant of the palpebral fissures, micro-cephaly, shortening of the metacarpals, metatarsals and/or phalanges were noted 3 times. Ptosis was striking in *Case 1,* who also had carp mouth, and these findings were also present in the case of Linarelli et al (4). Micrognathia, epicanthal folds and simian creases were noted twice (3, 4), though neither of our patients showed these features (Table 1). Except for strabismus and pre-sumed cortical blindness, the eyes were not abnormal. No palatal, GI, genital or urinary tract abnormalities have been shown.

Other than the mental and growth retardation, the physical features are quite distinct from those seen in a young boy with another type of deletion of chromosome 11 (technical deletion of 11p): aniridia, deafness, hypospadias and urinary tract malformations (V. M. Riccardi, unpublished observation).

A terminal deletion of 11q is the unifying feature among these patients. In 4 of them the break point is at q(23) and in one at q(21); it is of interest that there was no apparent proportionate increase in the severity of findings with the larger deletion. In one case the father was shown to have a balanced reciprocal trans-location [t(11;21) (q23;q22)] and the rearrangement was present in balanced and unbalanced forms in 4 generations (2). It is thus important to study parents and other family members. In our 2 cases only the mother of *Case 1* has been studied to date, although efforts to study the father of *Case 1* and both parents of *Case 2* are underway.

SUMMARY

A terminal deletion of the long arm of chromosome 11 is compatible with life, and the affected child uniformly shows growth and mental retardation and tele-canthus. However, no clear-cut, distinctive clinical syndrome is apparent. Mental and growth retardation alone may be the sole indications for chromosome analysis, as the virtual absence of dysmorphic features in some cases *(Case 2)* is noteworthy. Other appropriate family members should be evaluated to de-termine whether there is a familial balanced reciprocal translocation present and proper genetic counseling provided.

REFERENCES

1. Lewandoski, R. C., Jr. and Yunis, J. J.: New chromosomal syndromes. Am. J. Dis. Child. 129:515–529, 1975.
2. Jacobsen, P., Hauge, M., Henningsen, K. et al: An (11, 21) translocation in four genera-tions with chromosome 11 abnormalities in the offspring. Hum. Hered. 23:568–585, 1973.
3. Faust, J., Vogel, W. and Loring, B.: A case with 46,XX,del(11) (q21). Clin. Genet. 6:90–97, 1974.
4. Linarelli, L. G., Pai, K. G., Pan, S. F. and Rubin, H. M.: Anomalies associated with partial deletion of long arm of chromosome 11. J. Pediatr. 86:750–752, 1975.

Terminal (1)(q43) Long-Arm Deletion of Chromosome No. 1 in a Three-Year-Old Female*

Carl B. Mankinen, PhD, Joseph W. Sears, MD and Victor R. Alvarez, MD

With the recent advent of various chromosomal banding techniques, it is now possible to visualize minor structural changes in chromosomes which were heretofore undetectable. The present case is an excellent example of the need for employing these techniques whenever a chromosomal etiology is suspected. To the authors' knowledge, this is the first patient to be reported with a 46,XX,del (1) (q43) karyotype. A similar case (1) of a patient with a 46,XY,del(1) (q42) karyotype will be compared to the present study.

CASE REPORT

Our patient is a 3-year-old female of Latin-American extraction. Her mother was 15 years old at the time of her birth which followed a 40-week uncomplicated gestation. The mother had no known medications, drugs, x rays or infectious diseases throughout the pregnancy. Labor was spontaneous and a 2396 gm meconium-stained premature infant was delivered with an Apgar score of 5. She was microcephalic with an initial circumference of 32 cm. Persistent microcephaly and brachycephaly has characterized the child's head growth which remains below the 3rd percentile.

Physical examination at 2½ years revealed the following abnormalities: an interpupillary distance in the 75th percentile, epicanthus, moderate synophrys, bilateral convergent strabismus, short bulbar nose, mild prognathism, somewhat enlarged ears which are slightly posteriorly rotated with a flattened left pinna and sparse hair of a fine texture (Fig. 1). Examination of the hands indicated a mild soft tissue syndactylism of the 2nd and 3rd fingers, whorls present on all 10 digits and a slightly tapered shape to the fingers. Recent orthopedic evaluation of her bilateral pes planus revealed subluxation of the peroneal tendons around the fibula with severe pronation and heel valgus deformity. The feet

*Supported in part by a grant from the Moody Foundation of Galveston, Texas.

Birth Defects: Original Article Series, Volume XII, Number 5, pages 131–136
© 1976 The National Foundation

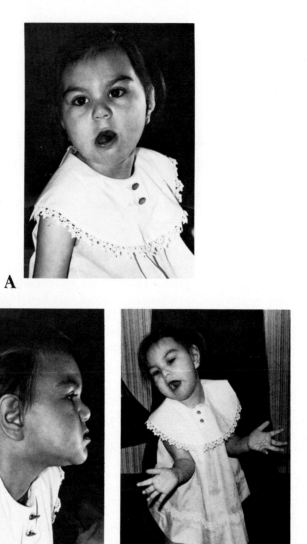

Fig. 1. Proposita, age 2½ years. A) Epicanthi, prominent frenulum, flattened left pinna and prominent mandible. B) Sparse hair, bulbar nose, prognathism and a slightly posteriorly rotated ear with attached lobe and prominent anthelix. C) Characteristic semiflexed elbows and partially hyperextended wrists with bent fingers. The head is usually inclined to the side.

exhibit asymmetry with the right foot being 12.5 cm long, 5 cm wide; and the left foot, 13.3 cm long and 6.4 cm wide. The patient keeps both elbows in a semiflexed position with both wrists partially hyperextended almost all the time.

Neurologic findings include a persistent absent pincer grasp, negative Babinski sign, slightly increased deep tendon reflexes and a mild tonic neck reflex. Grand mal seizures are documented from 9 months of age and are controlled with anticonvulsants. Coincidentally, convulsions were almost always associated with severe kidney infections.

Psychologic testing performed at 15 months of age on the Vineland Social Maturity Scale indicated a social quotient (SQ) of 43 and a social age (SA) of 6 months. The child has a persistent, unusual, high-pitched, shrill cry and speaks only 2 words at 3 years of age. She is able to sit upright only if placed in a vertical position and can retain this position for only a short period. She can stand briefly with support.

Subsequent examinations, occasioned by recurrent urinary tract infections, revealed the existence of vesicoureteral reflux, vaginal stenosis and a solitary left kidney. A ureterocystostomy and nephrostomy were performed at 33 months to correct the reflux. Clinical chemistry results at the time of surgery indicate that all SMA 12 values are in the normal range as are T_3, T_4, and serum cholesterol levels.

CYTOGENETIC STUDIES

Chromosomal analyses were conducted on peripheral blood lymphocyte cultures of the patient and her mother after the method of Moorhead et al (2). Replicate cultures were established in McCoy 5A and RPMI 1640 media (GIBCO) to minimize the possibility of aneusomic cell line selection. Slide preparations were stained using a modification of Seabright's Giemsa-trypsin technique (3). A minimum of 60 banded metaphase spreads on the patient and 30 on the patient's mother were photomicrographed and carefully analyzed. The father was unavailable for this study. Five of the patient's G-banded cells were fully karyotyped. Each demonstrated a deletion of the terminal, intensely stained bands (q43 and q44) of the long arm of one of the No. 1 chromosomes. A complete karyotype of the patient is shown in Figure 2. A series of homologous No. 1 chromosome pairs are shown somewhat enlarged in Figure 3.

The missing distal segment appears to contain less than 5% of the genetic material of the intact normal chromosome. As there was no evidence of a reciprocal translocation in any of the patient's karyotypes, it would appear that she is functionally monosomic for genes carried in the region of the deleted segment. Since there was no evidence of any secondary cell lines or mosaicism, the patient's karyotype formula is 46,XX,del(1) (q43) using the Paris Conference (1971) nomenclature (4). Examination of the mother's karyotype revealed no abnormalities of the No. 1 chromosomes.

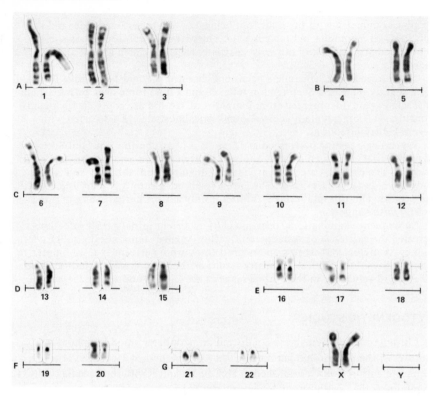

Fig. 2. Karyotype of the patient (G-banding with Leishman stain) showing normal chromosomal complement except for missing terminal band on the long arm of chromosome No. 1.

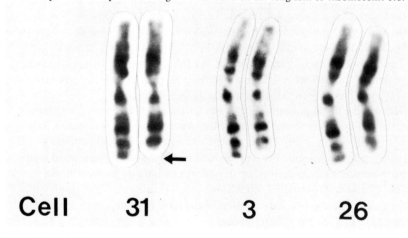

Fig. 3. G-banded No. 1 chromosome pairs from 3 other metaphase spreads. In each instance the abnormal homologue is to the right.

CYTOGENETIC DISCUSSION

Although chromosome No. 1 is the largest of the human chromosomes, relatively few studies have been reported of patients with partial deletions of this chromosome. Those reported are often of the ring chromosome type in which the exact location and identification of the deleted segment(s) has not been made. Only the one case to be reported by Koivisto et al is reasonably comparable cytogenetically (1). In their case 3 or possibly 4 distal bands are missing. Their patient's karyotype is 46,XY,del(1) (q42). He is a 4½-year-old boy followed from birth, whose head circumference and birthweight are comparable to those of our patient. Other similarities in their patient include limitation of the movements of both ankles and hips. He is severely retarded mentally and at age 4 years has not yet begun to use words. His length, however, is somewhat less and may be related to primary hypothyroidism and the growth hormone deficiencies which were found.

In general, patients with a partial deletion of the long arm of chromosome No. 1 have been microcephalic, of short stature, and have exhibited mental and motor retardation. As more clinical cytogenetic laboratories incorporate banding techniques into their routine protocols, additional patients with various deletions of the No. 1 chromosome will no doubt be detected. This will then permit more complete characterization of the phenotypic manifestations of the different chromosome No. 1 deletion syndromes.

SUMMARY

A 3-year-old Latin female is reported with a terminal deletion of the No. 1 chromosome, karyotype formula 46,XX,del(1) (q43). Principle clinical features include:

Anatomic – microcephaly; bilateral, convergent strabismus; epicanthus; brachycephaly; bulbar nose; sparse hair; partial soft tissue syndactylism between 2nd and 3rd fingers which are slightly tapered; whorls on all 10 fingers; mild prognathism; solitary kidney; vaginal stenosis; vesicoureteral reflux; asymmetric feet; and subluxation of peroneal tendons around the fibula with severe pronation and heal valgus deformity.

Neurologic – moderate motor and mental retardation; high-pitched, shrill cry; absent pincer grasp at 3 years; and grand mal seizures documented from 9 months of age.

ACKNOWLEDGMENTS

The authors express appreciation to Mr. D. Hale, Ms. M. Garcia, M. T. (ASCP), Ms. B. De La Houssaye and Mr. J. Tsai, M. S. for their technical assistance.

REFERENCES

1. Koivisto, M., de la Chapelle, A., Akerblom, H. K. and Remes, M.: A child with deletion of chromosome No. 1:46,XY,del(1)(q42), hypothyroidism and growth hormone deficiency. (Submitted for publication.)
2. Moorhead, P. S., Nowell, P. C., Mellman, W. J. et al: Chromosome preparations of leukocytes cultured from human peripheral blood. Exp. Cell Res. 20:613–616, 1960.
3. Seabright, M.: A rapid banding technique for human chromosomes. Lancet 2:971, 1971.
4. Bergsma, D. (ed.): "Paris Conference (1971): Standardization in Human Cytogenetics." Birth Defects: Orig. Art. Ser., vol. VIII, no. 7. White Plains: The National Foundation – March of Dimes, 1972.

X Short-Arm Deletion Gonadal Dysgenesis in Two Sibs

John R. Davis, MD, M. Wayne Heine, MD, Raymond F. Graap, MD, Elmer S. Lightner, MD and Harlan R. Giles, MD

The most common chromosomal basis for gonadal dysgenesis is a missing X chromosome, ie an X0 karyotype. Although rare, structural abnormalities of the X chromosome itself have been described which were associated with the clinical picture of gonadal dysgenesis, most commonly isochromosomes of X and translocations involving the X chromosome.

Prior to the availability of banding analysis, the nature of the translocations were merely inferred on the basis of clinical effect. Most patients evidenced clinical stigmata of gonadal dysgenesis. The vast majority involved autoradiography and were not confirmed by banding analysis. Translocations to the D group or to the X homologues account for 17 of the cases, while translocation to the E group has been reported only once and was not verified by banding. Although 4 authors report more than one affected family member, none of these reports include a balanced translocation carrier and the unbalanced consequence in the offspring.

CASE REPORTS

Case 1

This 16-year-old Mexican-American female was seen because of primary amenorrhea and short stature. Her history was unremarkable except for mild hypothyroidism. Physical examination revealed an immature, short, slightly obese girl whose weight was 43 kg and height was 132 cm. Widely spaced nipples were present on a shield chest. There was no breast development nor axillary hair. The escutcheon was sparse and the clitoris was small. Internal genitalia were felt to be infantile. Laboratory studies revealed spina bifida by x-ray examination, a bone age of 12 years, moderate hypothyroidism and Barr body counts of 5, 11 and 7%.

Birth Defects: Original Article Series, Volume XII, Number 5, pages 137–138
© 1976 The National Foundation

The chromosome banding from this patient revealed a short-arm deletion of the X chromosome, almost to the centromere. There was a single cell line, with no evidence of mosaicism.

Case 2

The 12-year-old sister of *Case 1* was short, immature, but normally proportioned (height 121.25 cm and weight 23.6 kg). Her physical examination revealed wide spacing of the nipples and valgus carrying angle of the arms. The genitalia were infantile and there was no pubic or axillary hair. She too was found to be significantly hypothyroid.

Her karyotype was identical to her sister's, showing a short-arm deletion of an X chromosome.

Chromosome analysis was immediately performed on the mother of the 2 girls. An X-16 balanced translocation, as a single cell line, was clearly identified by banding. Almost the entire short arm of the X was involved and transferred onto one of the No. 16 chromosomes.

Additional family members were studied. Another sister was found to be a balanced translocation carrier. Two other sibs were normal. Unfortunately, the father was not available for study. Thus, of the mother's 5 children, 2 were normal, 2 were Xp– and 1 was a balanced carrier.

DISCUSSION

The high transmission rate in the family studied is unusual among human translocation situations. The 40% unbalanced transmission rate not only exceeds that expected by chance alone, but is more than twice the observed transmission rate in translocation Down syndrome (6–20%).

In summary, we have presented a family demonstrating short-arm deletion of the X chromosome as a consequence of X-16 balanced translocation in the mother. The 2 Xp– sisters exhibit clinical signs of gonadal dysgenesis, while the balanced carriers are phenotypically normal. To our knowledge, this represents the only case example of both the balanced carrier state for an X translocation and its genetic consequence in the offspring. The transmission rate is significantly higher than expected in other human translocation situations.

Trisomy C and Cystic Dysplasia of Kidneys, Liver and Pancreas

John D. Blair, MD

Full autosomal trisomy for a C-group chromosome has rarely been identified, even in spontaneously aborted fetuses (1,2). The 1966 Geneva Conference accepted 4 instances of trisomy C in nearly 800 abortuses; only one specimen with full trisomy C was identified by Arakaki and Waxman (2) in their cytogenetic study of 127 spontaneously aborted embryos. Juberg et al (3) reported the first living individual with a full trisomy C associated with bilateral renal dysplasia and other congenital anomalies. Two mentally retarded patients with trisomy 8 have been reported by Casperson et al (4) and a newborn child with multiple congenital anomalies and trisomy 9 was described by Feingold and Atkins (5) and Kurnick et al (6).

The subject of this report is a full-term male infant with trisomy C demonstrated in cultured skin fibroblasts who lived for 17 hours. The patient was originally studied at the Cytogenetics Laboratory and Department of Pathology of the University of Kentucky Medical Center, Lexington.

CASE REPORT

Clinical course. A 3000-gm infant was delivered breech to a 28-year-old gravida 2, para 2 white female after membranes had been ruptured for over 12 hours. The mother was admitted to the hospital in labor and had had no prenatal care. She denied taking any medicines or drugs during her pregnancy, which apparently was uneventful. The placenta was described as having a hematoma near the emergence of the cord. The baby was born asphyxiated with an Apgar score of 1 at 1 minute and heart rate of 20/minute. Resuscitation procedures were carried out in the delivery suite where the baby was noted to be "difficult to ambu." Although his heart rate went up to 130/minute, he remained cyanotic and his respirations were labored and gasping. Physical abnormalities noticed at this time included bilateral anterior chamber ocular hemorrhages, low-set and malformed ears, small palpebral fissures, micrognathia and abnormal palmar

Birth Defects: Original Article Series, Volume XII, Number 5, pages 139–149
© 1976 The National Foundation

flexion creases (Fig. 1). The head was 35.5 cm in circumference and a single tooth of the lower gingiva was loosened up and removed shortly after birth. Poor air exchange and diffuse rales persisted throughout the entire postnatal period. The liver extended 2 cm below the right costal margin. The abdomen was soft and the external genitalia were normal male. Hypotonia, poor cry, fair Moro and absent suck reflexes were noticed. The skin was cyanotic, rubbery hard and redundant over the limbs and around the short neck.

At 4 hours of age, the blood pH was 7.008, the pO_2 was 42 and the pCO_2 was 91. Chest roentgenograms demonstrated bilateral pneumothoraces and pneumomediastinum. Pleural drain tubes were inserted and the patient was kept on positive pressure ventilation with 100% oxygen. Severe respiratory distress with cyanosis and metabolic acidosis persisted until he expired 17 hours after birth.

Postmortem examination. In addition to the abnormalities noted during life, there was a limbus dermoid in the temporal pole of the right eye. The occiput was unusually prominent and the left ear was low set and posteriorly rotated (Fig. 2). The nose was short and bulbous with a flattened bridge and bilateral

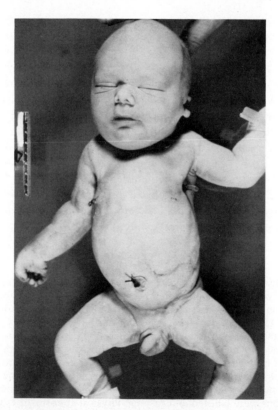

Fig. 1. General view at the time of postmortem examination. Note the bulbous nose and the broad philtrum.

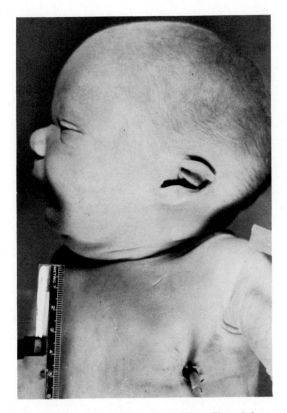

Fig. 2. Low-set malformed left ear. Peculiar facial profile and short neck.

notching of the ala. The neck was short. The left hand showed a single palmar flexion crease. The fingers were tapered and clinodactyly of the 5th fingers was present. The lungs were atelectatic and contained subpleural air blebs. The combined weight was 31 gm, as compared to an expected normal of 66 gm. Microscopic examination disclosed interstitial emphysema and diffusely nonaerated alveoli which appeared immature as they were lined by large cuboidal cells with clear-appearing cytoplasm; focal intraalveolar and interstitial anoxic hemorrhage and massive necrosis of the bronchial epithelium without inflammatory response were also noticed.

The kidneys (Fig. 3) were small (combined weight 11 gm, normal 29 gm), multicystic and, upon sectioning, no normal kidney parenchyma was evident. The pelves were present but they were very small and misshapen. The ureters were tiny and atretic. The urinary bladder was patent. Microscopically, renal dysplasia was present: severe distortion and absence of normal renal architecture with metaplastic cartilage embedded in dense fibrous and fibromyxomatous tissue. Multiple cysts and primitive ducts surrounded by concentric layers of fibromuscular tissue were present throughout (Fig. 4). No glomeruli could be

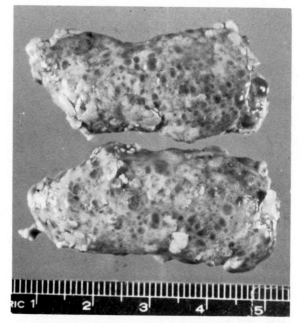

Fig. 3. Kidneys with multiple cysts visible on the external surface.

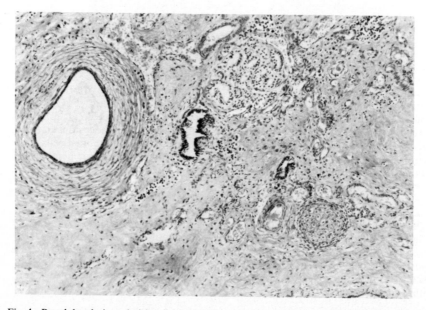

Fig. 4. Renal dysplasia: primitive ducts with fibromuscular collars and dense fibrous tissue. Note island of cartilage. (H and E, × 65)

identified. Although the gross appearance of the liver and pancreas was unrevealing, microscopic examination of these organs disclosed focal cystic dilatation of the interlobular ductal systems. Proliferation of biliary ductules and dysplastic interlobular ducts surrounded by collars of collagenous tissue were present in the portal tracts (Fig. 5). The pancreas was similarly affected with multiple cysts and cystic dilatation of the main duct which otherwise appeared empty and surrounded by fibrous tissue (Fig. 6). The brain showed diffuse subarachnoid hemorrhage and recent hemorrhage of the choroid plexus. There was no congenital anomaly of the central nervous system.

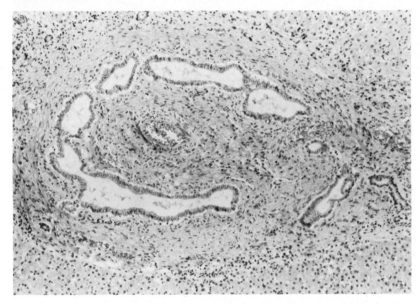

Fig. 5. Hepatic dysplasia: dilatation and proliferation of interlobular bile ducts surrounded by fibrous tissue. (H and E, ×65)

The spleen was grossly enlarged (weight 18 gm, normal 10 gm) and microscopically, the amount of hematopoiesis was in excess of the expected for a full-term infant. Testes and other organs were normal.

Cytogenetic studies. Skin was obtained 12 hours postmortem for cell culture and karyotype analysis and processed according to standard explant culture techniques. Cell growth was slow with harvesting of cells after 14 weeks. The number of metaphase plates suitable for analysis was limited. A total of 15 metaphase plates were examined: 12 had 47 chromosomes, the extra chromosome being one of the smaller C-group pairs; in addition, 2 of 3 cells with counts less than 47 were trisomic for a C-group chromosome although they were missing other autosomes. The karyotype of one cell with 46 chromosomes could not be established with certainty. A total of 8 metaphases were karyotyped (Table 1). Skin and other tissues obtained at autopsy were examined for X-heterochromatin. No Barr bodies were found in any of the tissues examined. Also, cultured skin fibroblasts were negative for X-heterochromatin.

Unfortunately, Giemsa-banding of the chromosomes of this child was not successful. However, careful matching of chromosomes suggests that the extra autosome is morphologically consistent with a No. 11 (Fig. 7).

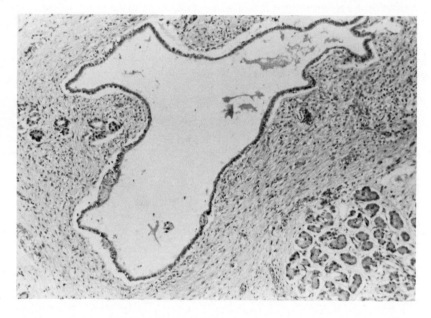

Fig. 6. Pancreas: cystic dilatation and proliferation of ductal system with fibrosis. (H and E, ×65)

TABLE 1. Chromosome Analysis

Specimen	Skin (Obtained 12 hours Postmortem)				
Chromosome count	45	46	47	48	TOTAL
Number of cells	1	2	12	0	15
Karyotypes	1	1	6	0	8
Interpretation	46,XY,+C(11?)				

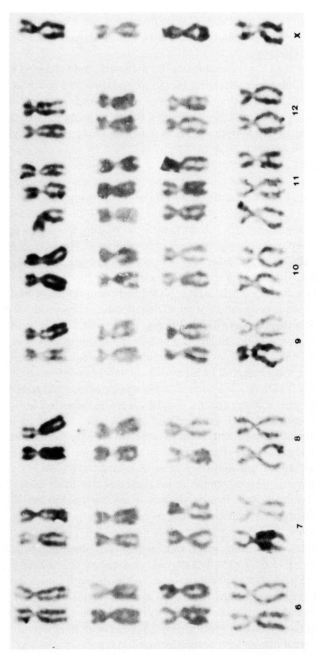

Fig. 7. Partial karyotype of 4 cells showing chromosomes 6–12 and X.

DISCUSSION

The documentation of trisomy C in cultured skin fibroblasts suggests an etiologic relationship between the chromosome anomaly and the renal, hepatic and pancreatic changes.

Full trisomy for a C-group autosome has previously been reported in individuals with multiple congenital anomalies (3–7); of these, the infant described by Juberg and co-workers (3) share a number of external and visceral anomalies with our patient, such as bilateral renal dysplasia, Potter facies, pulmonary hypoplasia, nasal configuration and hypotonia. The intrahepatic and pancreatic dysplasia were not present (8).

The association of trisomy C with renal and hepatopancreatic changes prompted a review of reported cases of trisomy or partial trisomy for chromosomes of group C in order to find possible occurrence of these visceral anomalies. A detailed pathologic study (6) of an infant with trisomy 9 does not indicate congenital cystic renal disease. Also, the 2 patients with full trisomy 8 reported by Casperson et al (4) do not present clinical evidence of renal disease. Similarly, the infant described by Kakati et al (7) does not have congenital cystic disease or dysplasia of the kidneys. Although full trisomy 11 has not been documented as yet, a number of patients with partial trisomy for this chromosome have been reported (9–14), but there is no mention of renal, hepatic and/or pancreatic dysplasia or cystic disease in these reports. The familial 11/13 translocation described by Rott et al (9) is of interest in that biliary atresia was mentioned as the cause of death of an infant of unknown karyotype with multiple congenital anomalies. In that case, the father was a balanced heterozygote. In another member of that kindred, aplasia of the left kidney was associated with myelomeningocele and congenital heart disease; cystic kidneys or renal dysplasia were not reported.

Trisomy C mosaicism with a normal cell line has been reported frequently (15–20). Some of these patients seem to represent trisomy 8 mosaicism syndrome (19,20), an anomaly characterized by ureterovesical reflux, hydronephrosis and genital, skeletal and dermatoglyphic abnormalities (4,21–24). Although there is no anatomic documentation of the renal status in those cases of trisomy 8 mosaicism syndrome with hydronephrosis, it appears that bilateral diffuse renal dysplasia of the type encountered in our patient is not clinically evident in the patients reported.

C-group autosomal trisomy involving chromosomes other than Nos. 8 and 11 has been reported (5,6,25–32). Renal cystic or dysplastic disease is not mentioned in these patients except for the infant with trisomy 9 mosaicism reported by Bowen et al (31) who stated that "the kidneys contained a number of microscopic cysts lined by epithelium." The observation is of interest but this brief description in the absence of specific details precludes an interpretation. No renal anomalies or renal disease have been mentioned in reports concerning mosaicism or partial trisomy for chromosome 9 (26–28) and chromosome 12 (29, 30). Renal cysts have been reported in trisomy 13 and trisomy 18 syndromes (33–36), but the type of renal malformation in these cases is the source of confusion and at the present, the evidence for a specific type of renal malformation in these syndromes is lacking (37), particularly in reference to trisomy 13.

Renal dysplasia, often classified as polycystic kidney type 2, has traditionally been regarded as sporadic and due to local disturbance of fetal development (38). Indeed, most cases of renal dysplasia probably reflect an acquired malformation. However, familial cases are being recognized with increasing frequency in a number of congenital malformation syndromes with variable patterns of autosomal inheritance. Hepatic and pancreatic cysts are known to coexist with the autosomal dominant type of polycystic kidney disease that is usually ascertained in adults. They also occur in the infantile form of polycystic kidney that is inherited as an autosomal recessive. Renal dysplasia, with or without cysts, is not known to be associated with hepatic and/or pancreatic cystic dysplasia except for the family described by Ivemark et al (39). The case reported by Juberg et al (3) with trisomy C and bilateral diffuse renal dysplasia apparently did not have the hepatopancreatic involvement found in the present case. Nevertheless, on the basis of phenotypic features, type of renal malformation and cytogenetic findings, it seems likely that the extra C-group autosome is etiologically responsible for the malformations and that the same chromosome is involved in both cases. Furthermore, the morphologic characteristics of the extra chromosome in this patient suggest trisomy 11.

The existing confusion regarding the morphologic classification of congenital cystic renal diseases and our incomplete knowledge of their pathogenesis and genetic aspects indicate the futility of further morphologic classifications and the need for a more complete genetic and cytogenetic investigation of patients with congenital dysplasias or cystic malformations of these organs.

SUMMARY

Trisomy C is documented in a 17-hour-old full-term male infant with bilateral diffuse renal dysplasia, Potter facies, pulmonary hypoplasia and other congenital anomalies. In addition to renal dysplasia, intrahepatic bile duct and pancreatic dysplasia are demonstrated microscopically.

The extra C-group autosome is morphologically consistent with a No. 11 and this is regarded as the etiology of the multiple malformations of this infant. The phenotype and the renal malformation are essentially similar to those described by Juberg et al (3) in an infant with trisomy C. It is suggested that the same chromosome is involved in both cases.

Karyotype analysis should be performed on patients suspected of congenital renal malformations since the evidence implicates trisomy C as the etiology of at least some cases of bilateral renal dysplasia.

ACKNOWLEDGMENTS

The author is grateful to Mrs. Carol Searcy and Mrs. Linda Smith for technical assistance and to Mrs. Jean Watters for typing the manuscript and helping with the bibliography.

REFERENCES

1. Geneva Conference: Standardization of procedures for chromosome studies in abortion. Cytogenetics 5:361–393, 1966.
2. Arakaki, D. T. and Waxman, S. H.: Chromosome abnormalities in early spontaneous abortion. J. Med. Genet. 7:118–124, 1970.
3. Juberg, R. C., Gilbert, E. F. and Salisburg, R. S.: Trisomy C in an infant with polycystic kidneys and other malformations. J. Pediatr. 76:598–603, 1970.
4. Casperson, T., Lindsten, J., Zech, L. et al: Four patients with trisomy 8 identified by the fluorescence and Giemsa banding techniques. J. Med. Genet. 9:1–7, 1972.
5. Feingold, M. and Atkins, L.: A case of trisomy 9. J. Med. Genet. 10:184–187, 1973.
6. Kurnick, J., Atkins, L., Feingold, M. et al: Trisomy 9: Predominance of cardiovascular, liver, brain and skeletal anomalies in the first diagnosed case. Hum. Pathol. 5:223–232, 1974.
7. Kakati, S., Nihill, M. and Sinha, A. K.: An attempt to establish trisomy 8 syndrome. Humangenetik 19:293–300, 1973.
8. Gilbert, E. F. and Juberg, R. C.: Personal communications.
9. Rott, H. D., Schwantitz, G., Grosse, K. P. et al: C11/D13 translocation in four generations. Humangenetik 14:300–305, 1972.
10. Sanchez, O., Yunis, J. J. and Escobar, J. I.: Partial trisomy 11 in a child resulting from a complex maternal rearrangement of chromosomes 11, 12 and 13. Humangenetik 22:59–65, 1974.
11. Tusques, J., Grislain, J. R., André, M.-J. et al: Trisomie partiele 11q identifiee grace a l'etude en "denaturation menagée" par la chaleur, de la translocation equilibree paternelle. Ann. Génét. 15:167–172, 1972.
12. Francke, U.: Quinacrine mustard fluorescence of human chromosomes: Characterization of unusual translocations. Am. J. Hum. Genet. 24:189–213, 1972.
13. Wright, Y. M., Clark, W. E. and Breg, W. R.: Craniorachischisis in a partially trisomic 11 fetus in a family with reproductive failure and a reciprocal translocation, t (6pt; 11q–). J. Med. Genet. 11:69–75, 1974.
14. Falk, R. E., Carrel, R. E., Valente, M. et al: Partial trisomy of chromosome 11: A case report. Am. J. Ment. Defic. 77:383–388, 1973.
15. Higurashi, M., Naganuma, M., Matsui, I. et al: Two cases of trisomy C6-12 mosaicism with multiple congenital malformations. J. Med. Genet. 6:429–434, 1969.
16. El-Alfi, O. S., Powell, H. C. and Biesele, J. J.: Possible trisomy in chromosome group 6–12 in a mentally retarded patient. Lancet 1:700–701, 1963.
17. Smith, D. W.: Autosomal abnormalities. Am. J. Obstet. Gynecol. 90:1055–1077, 1964.
18. DeGrouchy, J., Thieffry, S., Aicardi, J. et al: Trisomic partielle C par translocation t(Cq–; Dp+) et remaniement d'un C(p-q+). Arch. Fr. Pediatr. 24:859–868, 1967.

19. Monnett, P., Willemin-Clog, J., Ganthier, J. et al: La trisomie 6–12 (a propos d'une observation en mosaique). Arch. Fr. Pediatr. 24:869–879, 1967.
20. Riccardi, V. M., Atkins, L. and Holmes, L. B.: Absent patellae, mild mental retardation, skeletal and genitourinary anomalies and C-group autosomal mosaicism. J. Pediatr. 77:664–672, 1970.
21. Crandall, B. F., Bass, H. N. and Marcy, S. M.: The trisomy 8 syndrome: Two additional mosaic cases. J. Med. Genet. 11:393–398, 1974.
22. Malpuech, G., Dutrillaux, B., Fonck, Y. et al: Trisomie 8 en mosaique. Arch. Fr. Pediatr. 29:853–859, 1972.
23. Walravens, P. A., Greensher, A., Sparks, J. W. et al: Trisomy 8 mosaicism. Am. J. Dis. Child. 128:564–566, 1974.
24. Bijlsma, J. B., Wijffels, J. C. H. M. and Tegelaers, W. H. H.: C8 trisomy mosaicism syndrome. Helv. Paediatr. Acta. 27:281–298, 1972.
25. DeBault, L. E. and Halmi, K. A.: Familial trisomy 7 mosaicism. J. Med. Genet. 12:200–203, 1975.
26. Baccichetti, C. and Tenconi, R.: A new case of trisomy for the short arm of No. 9 chromosome. J. Med. Genet. 10:296–299, 1973.
27. Pedruch, P. E. and Weisskopf, B.: Trisomy for the short arms of chromosome 9 in two generations, with balanced translocations t(15p+; 9p-) in three generations. J. Pediatr. 85:92–95, 1974.
28. Haslan, R. H. A., Broske, S. P., Moore, C. M. et al: Trisomy 9 mosaicism with multiple congenital anomalies. J. Med. Genet. 10:180–184, 1973.
29. Uchida, I. A. and Lin, C. C.: Identification of partial 12 trisomy by quinacrine fluorescence. J. Pediatr. 82:269–272, 1973.
30. Hobolt, N., Jacobsen, P. and Mikkelsen, M.: Partial trisomy 12 in a mentally retarded boy and translocation (12; 21) in his mother. J. Med. Genet. 11: 299–303, 1975.
31. Bowen, P., Ying, K. L. and Chung, G. S. H.: Trisomy 9 mosaicism in a newborn infant with multiple malformations. J. Pediatr. 85:95–97, 1974.
32. Lewandowski, R. C. and Yunis, J. J.: New chromosomal syndromes. Am. J. Dis. Child. 129:515–529, 1975.
33. Warkany, J., Passarge, E. and Smith, L. B.: Congenital malformations in autosomal trisomy syndromes. Am. J. Dis. Child. 112:502–517, 1966.
34. Mottet, N. K. and Jensen, H.: The anomalous embryonic development associated with trisomy 13–15. Am. J. Clin. Pathol. 43:334–347, 1965.
35. Bartman, J. and Barraclough, G.: Cystic dysplasia of the kidneys studied by microdissection in a case of 13–15 trisomy. J. Pathol. 89:233–238, 1965.
36. Osathanondh, V. and Potter, E. L.: Pathogenesis of polycystic kidneys: Type 3 due to multiple abnormalities of development. Arch. Pathol. 77:485–501, 1964.
37. Bernstein, J. and Kissane, J. M.: Hereditary disorders of the kidney. In Rosenberg, H. S. and Bolande, R. P. (eds.): "Perspectives in Pediatric Pathology," Chicago: Year Book Medical Publishers, 1973, vol. 1, pp. 117–187.
38. Osathanondh, V. and Potter E. L.: Pathogenesis of polycystic kidneys. Type 2 due to inhibition of ampullary activity. Arch. Pathol. 77:474–484, 1964.
39. Ivemark, B. I., Oldfelt, V. and Zetterstrom, R.: Familial dysplasia of kidneys, liver and pancreas: A probably genetically determined syndrome. Acta Paediatr. Scand. 48:1–11, 1959.

9pter→p22 Deletion Syndrome: A Case Report

Penelope W. Allderdice, PhD,* Walter D. Heneghan, MB, BCh, FRCP (C)
and Emerita T. Felismino, MD

Alfi and associates have described a new chromosome deletion syndrome, based on homologies between the clinical features observed for 3 unrelated infants. Each was monosomic for the terminal segment of the short arm of chromosome No. 9 (1, 2). Their karyotypes were 46,del(9) (pter→p22). The case history of an additional male infant with the identical deletion and similar congenital anomalies is presented in this paper. The most striking facial feature of our case is trigonocephaly.

CASE REPORT

The proband was born to a 27-year-old mother and a 29-year-old father who were married in 1967. The mother had a miscarriage in 1967 in the second trimester. She suspected she had had 3 early first trimester spontaneous abortions, but only one was confirmed. In 1971 a normal male infant was born; subsequently the mother took birth control pills, these being discontinued in the month prior to conception of the proband. The mother was using a contraceptive foam during the month of conception. In the second to fifth months of this pregnancy, she took pills (of unspecified genre) to control nausea. At the end of the first trimester she presented with a history of contact with a rash, possibly rubella. Her rubella titre at the time was 1/10. Ten days later it was 1/40. Six weeks prior to delivery she was treated with penicillin for 4-5 days.

The proband was born by spontaneous vertex delivery at 41 weeks' gestation and weighed 3941 gm, with a head circumference of 37.5 cm. The Apgar score was 10 at 1 minute. The infant was immediately transferred to the Janeway Child

*Aided by a Basil O'Connor Starter Research grant from The National Foundation-March of Dimes.

Birth Defects: Original Article Series, Volume XII, Number 5, pages 151–155
© 1976 The National Foundation

Health Centre, where he was treated for hyperbilirubinemia, bilateral pneumothorax, and pneumonia while his multiple congenital anomalies were evaluated.
In the 8 months following his discharge, he has been hospitalized 6 times with
pneumonia, which is controlled with antibiotics. At the age of 9 months, he still
swallows poorly and often chokes. He is functioning at about 3 months in the
motor area and about 1 to 1½ months in the personal, social and language area.
He has bilateral abducens paralysis and nystagmus. He does not put weight on his
legs or try to crawl, and he cannot sit alone. The frontal and lateral views of the
proband are shown in Figure 1.

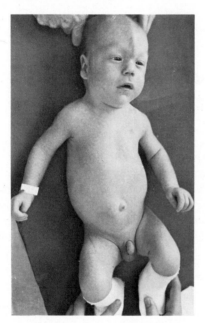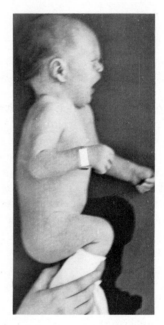

Fig. 1. Proband.

Congenital anomalies recorded at birth included downy facial hair; trigonocephaly with a prominent metopic suture; slight facial asymmetry with prominence of the left side; hypertelorism; mongoloid slant to the eyes; slight microphthalmia; a broad, depressed nasal bridge with a short anteverted nose; long
philtrum; and slight micrognathia. The ears were low set with abnormal auricular
lobules and extremely narrow external auditory canals. The palate was high
arched and the cry was noted as "high pitched." The short neck was slightly
webbed and the skin at the back of the neck was in loose folds. The nipples were
extremely wide spaced, the sternum was slightly depressed and the heart had a
grade 2/6 systolic murmur. The liver was palpable 1cm below the right costal
margin; the spleen and kidneys were not palpable. At 2 weeks there was no hepa-
tomegaly. The external genitalia were normal and the testicles were descended.
Both thumbs were clenched and there was bilateral campodactyly. The fingers

were long with broad fingernails and there were bilateral simian creases. The joints were generally stiffer than normal, the arms could not be lifted above a horizontal line. There was limited abduction of the hips. There were bilateral calcaneovarus deformities; however, the talonavicular joints were not dislocated and he did not have true rocker-bottom feet.

At 10 months the levels of serum immunoglobulins were: IgA 4 mg, IgG 220 mg and IgM 34 mg.

X-ray reports. Pelvic x-ray films showed that the upper femoral epiphyses were present and normally placed in the acetabula. There was no evidence of hip dysplasia. X-ray films of the skull showed an unusual pear-shaped outline of the supralateral margins of the orbits. During a recent review of the IVP film, the scapulas were noted as unusual, and the glenoid fossae were observed as shallow. At 10 months, abnormalities of the ribs as in pseudo-Hurler syndrome were noted by the radiologist. IVP showed the kidneys and bladder outlines to be normal.

EEG at 18 days showed the frequency of cortical rhythms to be greater than expected for the age of the infant; the significance was not known. There was no evidence at this time, nor at 10 months, to suggest epileptiform abnormality.

Cytogenetic Studies

The quinacrine fluorescent karyotype of the proband (Fig. 2, II_2) was 46,XY, del(9) (pter→p22) in 20 cells (Fig. 3).

The mother and father of the proband had normal quinacrine fluorescent karyotypes of 46,XX and 46,XY, respectively. There was no cytogenetic explanation for the poor reproductive history. In Figure 2, their chromosomes 9 are compared with the normal chromosome 9 and the deleted chromosome 9 carried by the proband.

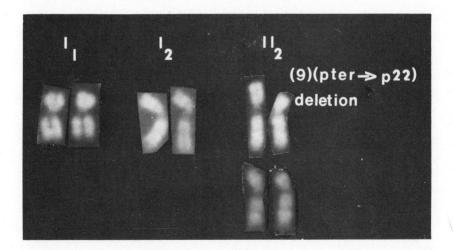

Fig. 2. Partial karyotypes comparing the chromosomes 9 from the father of the proband (I_1), the mother (I_2) and the proband (II_2).

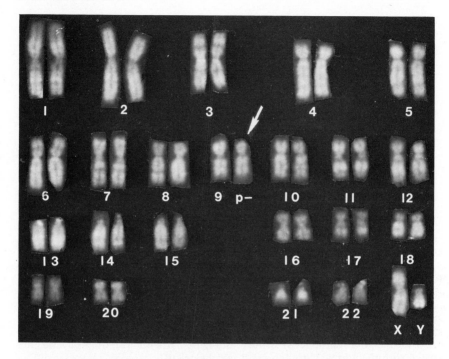

Fig. 3. Quinacrine fluorescent karyotype of the proband: 46,XY,del(9)(pter→p22).

DISCUSSION

Prior to the availability of the chromosome banding techniques, a child with the karyotype 46,XX,Cr and facies similar to our proband was described by Gacs and associates (3). An additional case with a C-group short arm deletion has been reported (4), but it is impossible to establish certain homology between banded and nonbanded karyotypes.

The proband is the fourth infant to be identified as carrying a deleted chromosome 9 short arm. In the presence of deletion (9) (pter→p22) these infants display a unique combination of anomalies (Table 1). The facies include trigonocephaly (with a prominent metopic suture in 2 of the 4); a broad, flat nasal bridge and anteverted nose; long upperlip; slight micrognathia; and a short neck with some webbing or loose skin folds at the back. Each had a cardiac murmur. Limited limb abduction was not noted for the 3 infants reported previously (1, 2); however, it is not possible to raise the arms of our proband above shoulder level. Unusual scalpulas and glenoid fossae are suggested on the IVP film from our proband at 3 weeks of age. Review of the x-ray films suggests the possibility of a lipid storage disease.

TABLE 1. Clinical Features Noted for Children With (9)(pter→p22) Deletion

	Our Proband	L. C.[2]	A. P.[1]	E. M.[1]
Sex	M	F	F	M
Birthweight (percentile)	97	30	10	90
Birth length (percentile)	97	95	90	50
Head circumference (percentile)	97	3	10	30
Age, mother	27	27	22	25
Age, father	29	24	25	29
Prominent forehead	+	+	+	+
Trigonocephaly	+	+	+	+
Mongoloid slant	+	+	+	−
Epicanthal fold	−	+	+	−
Hypertelorism	+	+	+	+
Flat nasal bridge	+	+	+	+
Anteverted nostrils	+	+	+	+
Long nasolabial distance	+	+	+	+
Low-set ears	+	+	+	+
Arched palate	+	+	+	+
Micrognathia	slight	slight	+	slight
Short neck	+	+	+	+
Webbed neck	+	−	−	+
Wide-set nipples	+	+	+	+
Cardiac murmur	+	+	+	+
Omphalocele	−	−	+	−
Long fingers	+	+	+	+
Broad nails	+	+	+	−
Hypertonia	+	−	+	+

REFERENCES

1. Alfi, O. S., Donnell, G. N., Crandall, B. F. et al: Deletion of the short arm of chromosome #9 (46, 9p-): A new deletion syndrome. Ann. Genet. (Paris) 16:17, 1973.
2. Alfi, O. S., Sanger, R. G., Sweeny, A. E. and Donnell, G. N.: 46,del(9)(22:): A new deletion syndrome. In Bergsma, D. (ed.): "Clinical Cytogenetics and Genetics," Birth Defects: Orig. Art. Ser., vol. X, no. 8. Miami: Symposia Specialists for The National Foundation-March of Dimes, 1974, pp. 27-34.
3. Gacs, G., Schuler, D. and Sellyei, M.: Familial occurrence of congenital malformations and ring chromosome (46,XX,Cr). J. Med. Genet. 7:177, 1970.
4. Elliott, D., Thomas, G. H., Condron, C. J. et al: C group chromosome abnormality (?10p-). Am. J. Dis. Child. 119:72, 1970.

The 9p−Syndrome*

Omar S. Alfi, MD, George N. Donnell, MD and Anna Derencsenyi

We have previously reported 3 infants (1,2) with partial deletion of the short arm of chromosome No. 9 distal to band 9p22 (Fig. 1). All 3 infants had several clinical features in common suggesting a relationship to the chromosomal defect.

Since then we have identified 2 new patients with the same chromosomal deletion. The clinical features in them were similar to those in the earlier cases. The diagnosis in these 2 patients was suggested on clinical grounds and later confirmed by cytogenetic studies.

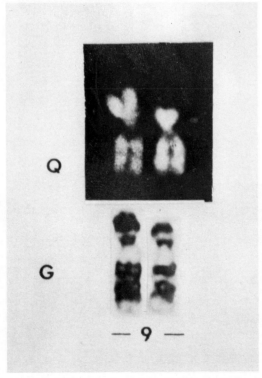

Fig. 1. Example of the 9p− chromosome by Q− and G−band staining.

*Supported in part by HEW Maternal and Child Health Service Project 422 and by the Michael J. Connell Foundation, Medical Genetics Fund.

Birth Defects: Original Article Series, Volume XII, Number 5, pages 157−160

MAIN CLINICAL FEATURES

The disorder constitutes a clinically identifiable syndrome (Fig. 2) consisting of mental retardation, sociable personality, trigonocephaly, mongoloid slant of the eyes, wide flat nasal bridge, anteverted nostrils, long upper lip, short neck, long digits mostly secondary to long middle phalanges and predominance of whorls on fingers (Table 1).

DISCUSSION

On comparing these patients to those with Tri 9p, (3,4) the "type et contre-type" concept of Lejeune (5) is suggested. The prominent forehead, mild exophthalmos, mongoloid eyes, anteverted nostrils and long philtrum seen in the 9p– patients contrast with the flattened forehead, mild enophthalmos, antimongoloid

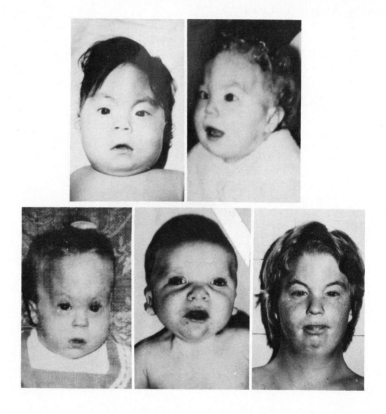

Fig. 2. Frontal views of the 5 patients.

TABLE 1. The Main Clinical Features in 5 Patients With 9p–

	Case 1	Case 2	Case 3	Case 4	Case 5
Sex	F	M	F	F	F
Age at diagnosis	21 mo	6 mo	11 mo	10 yr	20 yr
Birthweight percentile	10	90	30	50	50
Birthheight percentile	90	50	95	75	
Head circumference	10	30	3		
Age of mother	22	25	27	45	31
Age of father	25	29	24	48	34
DQ or IQ	35	30	40	44	45
Trigonocephaly	+	+	+	+	+
Mongoloid eyes	+	–	+	+	+
Epicanthal folds	+	–	+	+	+
Flat nasal bridge	+	+	+	+	+
Anteverted nostrils	+	+	+	+	+
Long philtrum	+	+	+	+	+
Low set ears	+	+	+	+	+
Abnormal ear lobule	±	±	+	+	+
High-arched palate	+	+	+	+	+
Micrognathia	+	+	+	+	+
Low hairline	±	+	+	+	+
Short neck	+	+	+	+	+
Webbed neck	–	+	–	+	+
Wide-set nipples	+	+	+	+	+
Cardiac murmur	+	+	+	–	–
Omphalocele	+	–	–	–	–
Long fingers	+	+	+	+	+
Square nails	+	+	+	+	+
Number of whorls	6	6	6	6	10

slant, retroverted nostrils and short philtrum in the Tri 9p. The contrast in the appearance of the hands is also interesting. In the 9p– patients the midphalanges are long, while in Tri 9p there is marked shortening of these phalanges.

From the number of 9p– patients seen in our laboratory, it may be assumed that the disorder is not very rare. However, cases might be missed because the deleted chromosomal segment is relatively small. In 2 of the 5 patients, the deletion was missed initially when the standard nonbanded chromosome technique was used and subsequently detected when banding was applied. It is hoped that the delineation of the constellation of clinical features in the 9p– patients will help in focusing attention on the associated small chromosome deletion.

REFERENCES

1. Alfi, O. S., Donnell, G. N., Grandall, B. F. et al: Deletion of the short arm of chromosome no. 9 (46,9p−): A new deletion syndrome. Ann. Genet. (Paris) 16:17, 1973.
2. Alfi, O. S., Sanger, R. G., Sweeny, A. E. and Donnell, G. N.: 46, del(9)(22:). A new deletion syndrome. In Bergsma, D. (ed.): "Clinical Cytogenetics and Genetics," Birth Defects: Orig. Art. Ser., vol. X, no. 8. Miami: Symposia Specialists for The National Foundation − March of Dimes, 1974, pp. 27−34.
3. Rethore, M. O., Larget-Piet, L., Abonyi, D. et al: Sur quatre cas de trisomie pour le bras court de chromosome 9. Individualisation d'une nouvelle entite morbide. Ann. Genet. (Paris) 13:217, 1970.
4. Rethore, M. O., Hoehn, H., Rott, H. D. et al: Analyse de la trisomie 9p par denaturation ménagée. Humangenetik 18:129, 1973.
5. Lejeune, J.: Types et contretypes. Journ. Parisiennes Pediatr., 75, 1966.

Partial Monosomy and Partial Trisomy for Different Segments of Chromosome 13 in Several Individuals of the Same Family*

Robert S. Wilroy, Jr., MD, Robert L. Summitt, MD, Paula Martens, BS, W. Manford Gooch, III, MD, Carol Hood, BA and Winfred Wiser, MD

At the 1974 Birth Defects Conference held at Newport Beach, California, we presented a family in whom 3 members had partial trisomy for chromosome 13. Two of the 3 had partial trisomy for the distal long arm of chromosome 13 and the third member was trisomic for the proximal portion of chromosome 13. We now wish to report an additional member of this family, who is partially monosomic for chromosome 13.

As shown in the pedigree (Fig. 1), female members are carriers of a balanced reciprocal translocation involving chromosomes 13 and 17, [46,XX,rcp(13;17) (q14;p13)]. One of these carriers became pregnant for the fifth time in October 1974. Amniocentesis was performed at 14 weeks and the fetal karyotype was interpreted as 46,XX,del(13)(q14) (Fig. 2). A therapeutic abortion via hysterotomy was performed and a malformed 420 gm female fetus delivered (Fig. 3). She had a single umbilical artery. Only 4 fingers were noted on each hand (Fig. 4) and only 4 toes on one foot (Fig. 5). The 2 lateral toes of the other foot were syndactylous. Radiographs confirmed the absence of the digits and also the absence of the appropriate metacarpals and metatarsals (Fig. 6).

Internal malformations were numerous, and included agenesis of the corpus callosum, a common aorticopulmonary trunk, an interventricular septal defect, and atresia of the first portion of the duodenum. The kidneys were extremely hypoplastic (Fig. 7) and the majority of renal tissue was medullary. All portions

*Supported in part by Special Project No. 900, Division of Health Services, MCHS, HSMHA, DHEW and a grant from The National Foundation — March of Dimes.

Birth Defects: Original Article Series, Volume XII, Number 5, pages 161–167
© 1976 The National Foundation

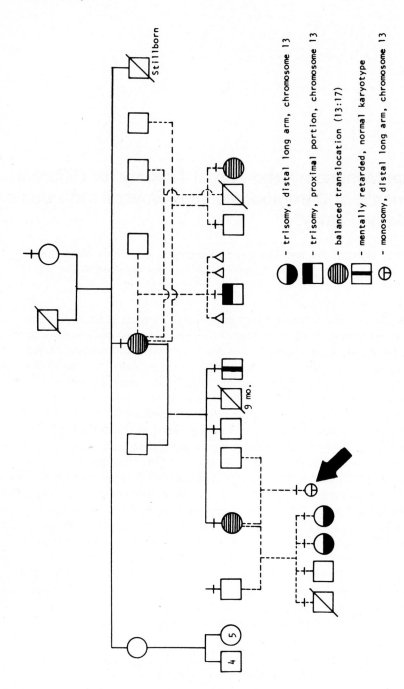

Fig. 1. Pedigree of family in whom members are partially trisomic for different segments of chromosome 13. The fetus (arrow) is partially monosomic for chromosome 13. A horizontal mark above each symbol denotes a chromosomal analysis has been performed.

- trisomy, distal long arm, chromosome 13
- trisomy, proximal portion, chromosome 13
- balanced translocation (13:17)
- mentally retarded, normal karyotype
- monosomy, distal long arm, chromosome 13

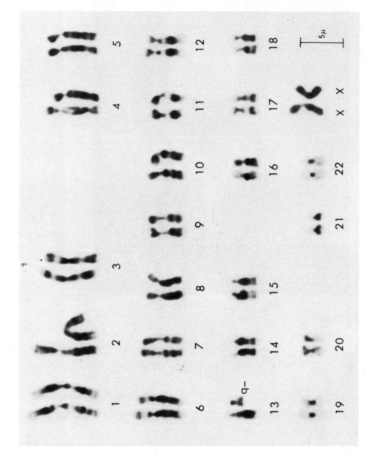

Fig. 2. Karyotype of cells from amniocentesis (IV-5 in the pedigree) 46,XX,del(13)(q14).

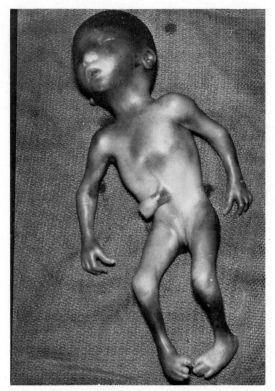

Fig. 3. Fetus, full body view.

of the renal tubular system were dilated (Fig. 8). The internal genitalia were normal. Examination of serial sections of the eyes revealed no evidence of a retinoblastoma.

Fibroblasts cultured from the fetal lung revealed the same karyotype as those cells obtained from amniocentesis, 46,XX,del(13)(q14) (Fig. 9).

The half sisters of this fetus are trisomic for the distal long arm of chromosome 13, mentally retarded, have long philtrums and 6 digits on each hand and foot. The maternal half uncle of the fetus, who is trisomic for the proximal portion of chromosome 13, is also mentally retarded, has a very short philtrum and 5 digits on each hand. The malformations seen in this fetus are more severe than those previously described in living children who have deletions of the long arm of chromosome 13.

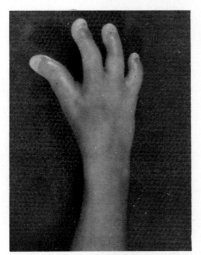

Fig. 4. Hand of fetus illustrating only 4 fingers.

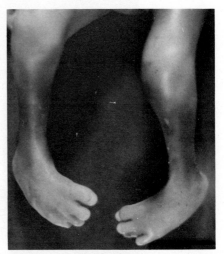

Fig. 5. Feet of fetus illustrating only 4 toes of one foot and 5 toes of the other foot of which the lateral 2 toes are syndactylous.

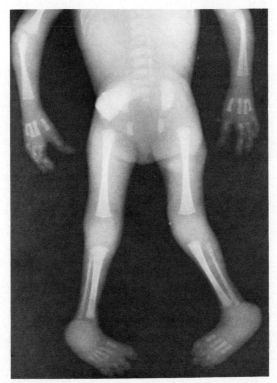

Fig. 6. Radiograph of the fetus illustrating absence of digits and absence of appropriate metacarpals and metatarsals.

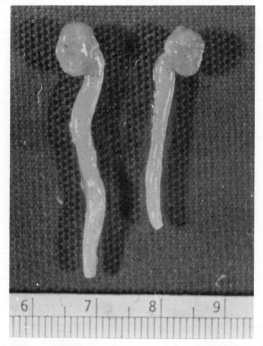

Fig. 7. Hypoplastic fetal kidneys.

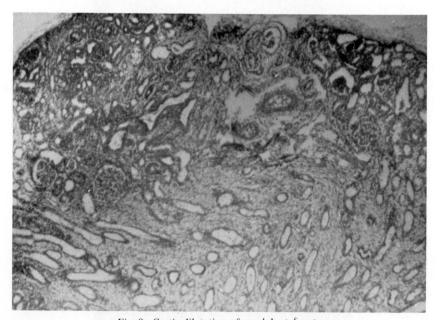

Fig. 8. Cystic dilatation of renal ductal system.

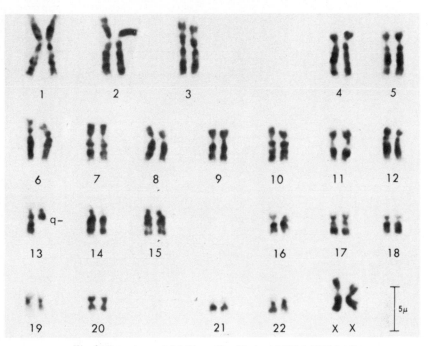

Fig. 9. Karyotype of fetal lung fibroblasts, 46,XX,del(13)(q14).

Ambiguous Genitalia and Mental Retardation Associated With a Translocation 46, XY, t(9; 10) and a Deletion in 9q*

Edmund C. Jenkins, PhD, R. S. K. Chaganti, PhD, Lorraine Wilbur, BS and James German, III, MD

In 1971, German and Simpson (1) described a male pseudohermaphrodite with ambiguous genitalia, mental retardation and a chromosome complement of 46,XY,t(Cq+; Cq−). We have made a cytogenetic reinvestigation of this patient utilizing G-, Q-, R-, and C-staining methods in order to identify more accurately the chromosome rearrangement present in his cells. We have found that chromosomes Nos. 9 and 10 were involved in the translocation; in addition, we have also been able to demonstrate a deletion of a band in the centromeric heterochromatin (9q12) of his maternal No. 9 chromosome.

CASE REPORT

A detailed description of the clinical and laboratory findings on the propositus has been given by German and Simpson (1). To summarize briefly, the propositus, born in 1968, weighed 3500 gm at birth. The genitalia were ambiguous and a clinical suspicion of adrenogenital syndrome was entertained because of vomiting and dehydration which developed soon after birth. He had a sloping forehead, a large bulbous nose and a small-appearing mandible. The unusually small (1.5 cm long) phallus was described by the attending physician as a "large clitoris" without a meatus. The labial-scrotal folds resembled labia. They were slightly pigmented and had a midline cleft. On the sixth day of life, testes of normal size and consistency were first palpated high in the "labia." At the age of 7½ years the endogenous urinary etiocholanolone/androsterone ratio was 1:4, which is within normal range (analysis by Drs. J. Imperato-McGinley and R. E. Peterson) (2).

At the present time the craniofacial dysmorphia, though still present, is less obvious than it was at birth (Fig. 1). The child is severely retarded. He does not speak and is unable to feed himself. In addition, he is hyperactive and displays bizarre mannerisms resembling choreiform movements. Extensive plastic surgery has been performed on his genitalia.

*Supported in part by NIH grants HD 04134 and HL 09011

Birth Defects: Original Article Series, Volume XII, Number 5, pages 169—173
© 1976 The National Foundation

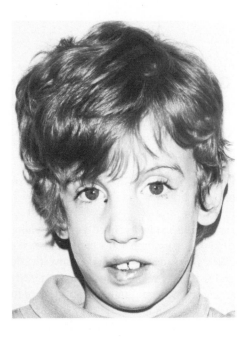

Fig. 1. Propositus, age 7 years.

A total of 251 cultured dermal fibroblasts and blood lymphocytes were studied at metaphase making use of G-, Q- or R-staining methods. In all of them a translocation involving the long arms of chromosomes 9 and 10 was present. The long arm of the aberrant No. 9 was much shorter than that of its normal homologue. In cells stained for G-banding, the secondary constriction region (which stains light) of the aberrant No. 9 appeared shorter than in the normal homologue. Distal to it were a short dark band and a short light band. In cells stained for C-banding, the C-band on the aberrant No. 9 was impressively small, clearly shorter than that in the normal homologue. Distal to it was a short lightly stained segment (Figs. 2 and 3). The long arm of the aberrant No. 10 was much longer than that of its normal homologue; the banding was normal from the centromere distally to band q25, at which point was attached a piece of chromosome exhibiting the banding pattern of 9q21→9qter (Fig. 2). We interpret the banding patterns of the aberrant chromosomes 9 and 10 as an indication of a reciprocal translocation.

By comparing measurements of the normal and the rearranged chromosomes 9 and 10 in 60 cells (20 with G-banding, 22 with R-banding and 18 with C-banding), we were able to demonstrate that the combined length of the long arms of the rearranged Nos. 9 and 10 (ie 9q− plus 10q+) was less by about 6%

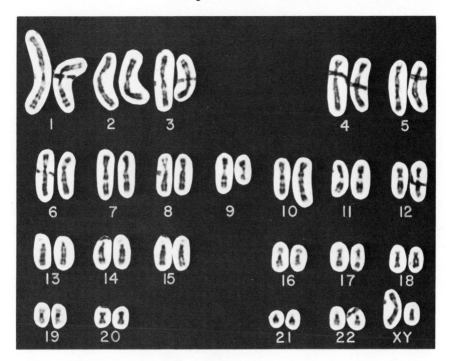

Fig. 2. G-banded lymphocyte in metaphase from the patient showing the translocation between chromosome Nos. 9 and 10.

than that of the normal 9 and 10 (ie 9q plus 10q). The C-band on the aberrant No. 9, as described above, was shorter than the one on the normal No. 9. This difference in C-band size on the No. 9 was responsible for the difference in length between the normal and rearranged chromosomes 9 and 10 mentioned above and could be due either to an inherited polymorphism or deletion of a segment of the C-band during the translocation. C-band patterns from the parents' No. 9 chromosomes were studied. Both the father's No. 9 chromosomes had conspicuous C-bands which were equal in size. One of the No. 9 chromosomes of the mother had a C-band which was of the same size as that on the father's No. 9, while the other No. 9 had a slightly longer C-band (Fig. 3). In the propositus the normal No. 9 had a C-band of the same size as that on his father's No. 9. We interpret the propositus' rearranged No. 9 (with the unusually short C-band) as a derivative of one of the parental No. 9 chromosomes in which deletion of part of 9q12 has taken place. Thus, at least 3 breaks appear to have occurred in the chromosome complement of the propositus in order to have given rise to his karyotype – 2 in No. 9 (one within band q12 and the other at band q13) and one in No. 10 (at band q25). The karyotype, in terms of the *Paris Conference* (1971) (3) nomenclature, can be written: 46,XY,del,t(9; 10) 9pter→9q12.5:: 10q25→10qter; 10pter→10q25::9q21→9qter).

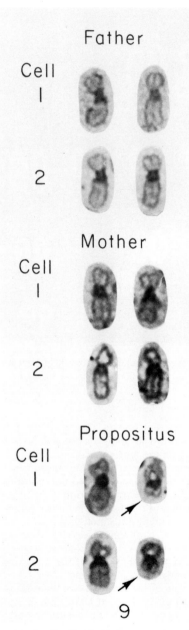

Fig. 3. Partial karyotypes showing the No. 9 chromosomes from 2 C-banded cells from the father, the mother and the patient. The arrow indicates the patient's aberrant No. 9 chromosome. (See text for details.)

DISCUSSION

The most interesting aspect of this study, which is also the reason for our updating the report on this patient, is the fact that it has been possible to demonstrate an autosomal deletion in association with a translocation in an individual in whom, except for severe mental retardation, genital ambiguity is the main abnormality. If these are causally related, then a genotypic component could be postulated to be present on either chromosome No. 9 or No. 10, which either through loss by deletion or through position effect would lead to abnormal genital development.

REFERENCES

1. German, J. and Simpson, J. L.: Abnormalities of human autosomes. 1. Ambiguous genitalia associated with a translocation 46,XY,t(Cq+; Cq−). In Bergsma, D. (ed.): Part X "The Endocrine System," Birth Defects: Orig. Art. Ser., vol. VII, no. 6 Baltimore: William & Wilkins Co. for The National Foundation − March of Dimes, 1971, pp. 145−149.
2. Imperato-McGinley, J., Guerrero, L., Gautier, T. et al: Steroid 5α-reductase deficiency in man. An inherited form of male pseudohermaphroditism. In Bergsma, D. (ed.): "Genetic Forms of Hypogonadism," Birth Defects: Orig. Art. Ser., vol. XI, no. 4. Miami: Symposia Specialists for The National Foundation − March of Dimes, 1975, pp. 91−103.
3. Bergsma, D. (ed.): "Paris Conference (1971): Standardization in Human Cytogenetics," Birth Defects: Orig. Art. Ser., vol. VIII, no. 7. White Plains: The National Foundation − March of Dimes, 1972.

Selected Abstracts

DERMATOGLYPHICS AND UNUSUAL ANORECTAL AND GENITAL ANOMALIES IN TRISOMY 8 MOSAICISM

J. D. Blair

Departments of Pathology and Pediatrics, Cardinal Glennon Memorial Hospital for Children and Saint Louis University School of Medicine, St. Louis, Missouri

A 2-month-old sex-chromatin negative male was first examined because of failure to thrive and multiple congenital anomalies. At 2 and 9 months of age, 46,XY/47,XY, +8 mosaicism was demonstrated in cultured peripheral blood leukocytes. The patient presented with anorectal and genital anomalies which have not been described in patients with trisomy 8 mosaicism: abnormally large penis, diffuse enlargement of the testes, a corrugated bifid scrotum and an ectopic anterior perineal anus at the base of the large scrotal sac. The anterior urethra and meatus were normally placed. There was no clinical evidence of sexual maturation or adrenocortical tumor.

The following dermatoglyphics were peculiar and characteristic: deep vertical furrows in the left palmar interdigital spaces and in I2, I3 and I4 areas of the right hand; single flexion creases of both thumbs and left 5th finger; unusual course and exit of D and T palmar lines with distal displacement of axial triradii and increased *atd* angle. There are patterns in the thenar-I1, I3 and I4 areas bilaterally as well as in the hallucal-I1, I2 and I3 areas of both feet. There are arches in all fingers except for ulnar loops in both 5th fingers and right 4th finger.

In addition, this patient presented most of the abnormalities previously described in this syndrome, including hypertonicity, asymmetric skull, abnormal ears, peculiar facies, absent patella and vesicoureteral reflex with bilateral hydronephrosis. Developmental retardation was evident at the age of 9 months at which time mental retardation was suspected.

Whether macrogenitosomia and anorectal malformations are part of this syndrome or whether they might be related to a possible structural abnormality of one of the three chromosome 8 of this patient remains an undecided issue.

Birth Defects: Original Article Series, Volume XII, Number 5, pages 175—177
© 1976 The National Foundation

VIABILITY OF XO FETUSES

A. S. Tucker, A. H. Chang and E. V. Perrin

Departments of Radiology and Pathology, University Hospitals of Cleveland and Case Western Reserve University, Cleveland, Ohio

Chromosome abnormalities are approximately 50 times as frequent in still-born fetuses as in live births. One of the most common anomalies is a sex-chromosome deficiency which is most frequently associated with Turner syndrome.

We have seen 3 stillborn fetuses in which the bodies were edematous and distinguished by particularly prominent fluid-distended sacs in the nuchal regions. Histologic examination demonstrated dilated lymphatics in many of the visceral structures. The posterior cervical sacs were lined with endothelium and were considered to represent hygromatous lymphatic cysts.

It has been suggested that these Nackenblasen are severe forms of the same deformity that leads to neck webbing in surviving children with the XO karyotype.

COMPARISON OF CHROMOSOME BREAKAGES IN LYMPHOCYTES AND FIBROBLASTS FROM CONTROL WOMEN AND WOMEN TAKING ORAL CONTRACEPTIVES

L. G. Littlefield and J. B. Mailhes

Medical Division, Oak Ridge Associated Universities, Oak Ridge, Tennessee

To determine whether exposure to synthetic hormones resulted in increased chromosome damage in cells other than lymphocytes, we evaluated series of lymphocyte and fibroblast cultures from 5 control women and 5 women taking oral contraceptives (OC). The results of this study showed 1) no differences in chromosome breakages between lymphocytes and fibroblasts; 2) no differences in breakages in replicate fibroblast cultures exposed to either fetal calf serum, autologous serum, homologous serum, or homologous serum from OC users; 3) no differences in breakages between lymphocyte and fibroblast cultures from OC users compared to lymphocyte and fibroblast cultures from control women and 4) no increase in the frequency of cytogenetically aberrant stem lines in fibroblast cultures from women taking OC. These findings suggest that synthetic hormones do not cause increased chromosome breakages in fibroblasts or lymphocytes from women taking OC.

Q-BANDING OF SINGLE CHROMATIDS

O. S. Alfi and T. N. Gardner

Department of Pediatrics, University of Southern California and Division of Medical
Genetics, Children's Hospital of Los Angeles, Los Angeles, California

The single chromatid staining technique, based on the quenching by 5 BUDR of the fluorescence caused by Hoechst #33258 stain, clearly visualizes the sister chromatid exchanges. A modification of that technique was employed, in which the single stained chromatid is Q-banded. The bands where the sister chromatid exchanges occur are being studied to determine whether certain sites are more susceptible to exchanges.

THE CLINICAL DIAGNOSIS AND MANAGEMENT OF 11 XYY MALES

F. A. Baughman Jr. and J. D. Pool

Sections of Neurology and Endocrinology, Blodgett Memorial Hospital and St. Mary's
Hospital, Grand Rapids, Michigan

Nine XYY males were ascertained in a private neurologic practice and 2 were ascertained by practitioner colleagues and referred for consultation. Five of 7 were significantly taller than their parents or sibs. Six of 11 had seizures in childhood and of the 6, 2 have "idiopathic" epilepsy and 4 have a paroxysmally abnormal EEG. In 1 of 11, essential tremor comprised the chief complaint while 4 had tremor of possible clinical significance. Seven of 11 have acne. Six have gynecomastia. Four of 11 have varicose veins and 2 of them have had recurrent pulmonary emboli. In all cases the carrying angles were below the average of $10.4°$ valgus of the XY control population. In 3 of 11 there was cubitus rectus, or $0°$. None had radioulnar synostosis or radial head dislocation.

Four of 11 were late or poor in walking and/or talking as compared to their sibs. Seven of 11 were subnormal in school performance as compared to their sibs. In 6 of the 11 the IQ determinations were lower than in their sibs. Five of 11 XYY males, but none of their sibs, had been incarcerated involuntarily either in relation to a criminal offense or psychiatric symptomatology.

The somatic and psychologic characteristics were variable but recurring. When met in aggregate, they suggest the presence of an XYY genotype.

SECTION III:
MALFORMATION SYNDROMES

Peters Anomaly With Pulmonary Hypoplasia

Marilyn J. Bull, MD and Jules L. Baum, MD

Peters anomaly (mesodermal dysgenesis of the cornea) is a rare developmental defect of the anterior segment of the eye. Major findings include congenital central corneal opacities, abnormalities of the posterior corneal stroma and defects of Descemet membrane (1, 2). Anterior synechiae often extend from the pupillary zone of the iris to the periphery of the corneal opacification. Glaucoma has been reported in over 50% of the cases. Sclerocornea, anterior polar cataracts, microphthalmia and persistent pupillary membrane may be present (1–3).

Iris abnormalities occur rarely in Peters anomaly while Rieger anomaly (mesodermal dysgenesis of the iris) is characterized by generalized hypoplasia of the iris stroma (1). Many similarities exist between these 2 conditions but only one example of a combined lesion incorporating the findings of Rieger mesodermal dysgenesis of the iris and central leukoma with defective Descemet membrane and anterior synechiae has been confirmed by histopathologic evaluation (4).

In addition to involving developmental defects of the eye, Peters anomaly has been described in association with several systemic abnormalities. Alkemade described associated anomalies of the skull, face, limbs, and central nervous, cardiovascular, GU and digestive systems (1). Findings associated with Peters anomaly reported by Townsend et al included hyaline membrane disease, mental retardation, congenital cardiac abnormalities, low birthweight and Lowe syndrome (4). Two additional infants in this report were noted at autopsy to have multiple abnormalities compatible with, but not proved of, trisomy 13. Another infant died at 1 hour of age and autopsy examination revealed pulmonary and intracranial hemorrhage compatible with intrauterine anoxia.

CASE REPORT

Our patient, born with bilateral opaque central corneas and pulmonary hypo-

Birth Defects: Original Article Series, Volume XII, Number 5, pages 181–186
© 1976 The National Foundation

plasia, was the first child of a 23-year-old mother and 33-year-old father. The gestation was 36 weeks and complicated by maternal emesis throughout the pregnancy. Edema in the third trimester was treated with a diuretic. Placental membranes ruptured 48 hours prior to delivery by C section. Birthweight was 2693 gm and the 1 minute Apgar was 5. Both parents are Portuguese from the Azore Islands but there was no known consanguinity. There was no family history of ophthalmologic problems or other birth defects.

On physical examination the patient was in the 50th percentile for height and weight and at the 25th percentile for head circumference. She was noted to have bilateral cloudy corneas, antimongoloid slant of the eyes, a narrow palatal vault and widely separated sagittal sutures (Figs. 1 and 2). The remainder of the physical examination was normal.

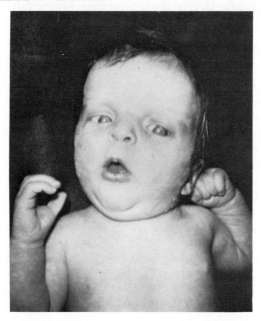

Fig. 1. Patient at age 4 weeks with broad forehead and antimongoloid slant of eyes.

Ocular examination at age 18 days revealed good photophobic response to a strong light stimulus in each eye. No optically elicited movements were invoked. The extraocular movements were grossly normal. The lids and conjunctivas in each eye were normal. Corneal diameter was 10.5 mm in each eye. Corneal sensitivity was normal using the Cochet-Bonnet anesthesiometer. A crescent-shaped opacity consistent with scleralization of the cornea was seen superiorly in each eye extending 2 mm inside the limbus concentrically and running from 9 o'clock to 3 o'clock (Fig. 3). Numerous superficial blood vessels were seen within these scleralized areas. A moderately dense stromal opacity, 7-8 mm in diameter, was seen centrally in each cornea. The opaque central stroma in each eye was approximately half the normal corneal thickness, the thinning resulting from a circum-scribed absence of the posterior half of the central stroma. Except for the scleralization of the superior cornea mentioned above, the peripheral corneas were normal. Several strands running from the iris to the central opacified cornea were seen in each eye. The pupils were 3 mm in diameter. Iris movement was seen when

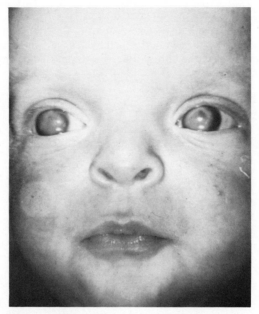

Fig. 2. Bilateral cloudy corneas.

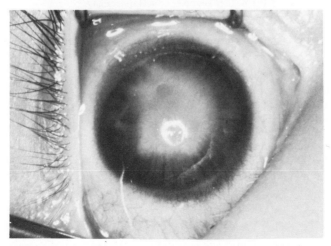

Fig. 3. Dense central stromal opacity and scleralization of superior cornea.

a light was shown in each eye but there was no change in the pupillary diameter. The posterior segments could not be visualized through the central opacities. At age 24 days contact A & B scan ultrasonography demonstrated no abnormality of the posterior segment in either eye.

Under general anesthesia at age 25 days, the intraocular pressure in each eye using the Perkins applanator was 17 mm Hg. Following this examination, an optical iridectomy was performed in the left eye and at that time the visible portions of the lens, vitreous and retina were seen to be grossly normal (Fig. 4). An optical iridectomy was performed in the right eye at age 2 months. A small area of the fundus seen through the optical iridectomy appeared normal following surgery.

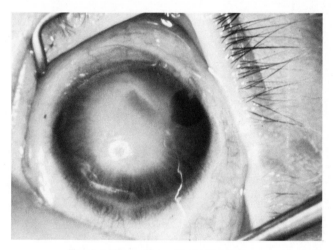

Fig. 4. Left eye after optical iridectomy.

Blood and urine amino acids, urine organic acids, urine for reducing substances, Berry spot test for mucopolysaccharides and the VDRL were normal. Serum IgM was 0.22 mg/cm³ at age 3 weeks. Chromosome analysis revealed 46,XX, a normal female karyotype. There were no intracranial calcifications on skull radiographs. Chest radiographs revealed a density in the right thorax consistent with hypoplasia of the right upper lobe and possibly a portion of the right middle lobe of the lung (Fig. 5). On chest fluoroscopy there was no evidence of a vascular ring or compression by abnormal structures.

At age 9 months the patient saw and reached for toys with ease. A horizontal gross pendular nystagmus was present and increased on lateral gaze. Her developmental milestones were normal and her general health had been good, with the exception of one hospitalization for bronchiolitis.

DISCUSSION

The developmental abnormalities of the anterior segment of the eye are difficult to classify. They are often designated by various eponyms and many cases

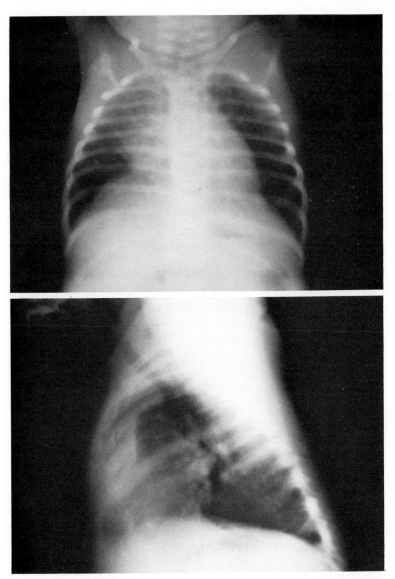

Fig. 5. Radiodensity in right thorax consistent with hypoplasia of right upper lobe.

do not fall easily into preestablished categories. Waring et al recently classified these disorders anatomically in an attempt to eliminate the confusion caused by eponymic designations (5).

The pathogenesis of Peters anomaly is controversial. Recently Townsend reviewed the literature and discussed the possible etiologies (6).

Peters anomaly may occur as an isolated finding or as one manifestation of a syndrome with systemic abnormalities. Autosomal recessive transmission has been suggested in several reported cases (7, 8) although a few reports are consistent with autosomal dominant inheritance (9). No consistent inheritance pattern has been established.

Pulmonary hypoplasia is a mass of poorly differentiated lung parenchyma connected to a malformed bronchus and is frequently associated with other congenital anomalies such as diaphragmatic hernia (10). Cardiac, GI and skeletal malformations have also been described with pulmonary hypoplasia. The high incidence of associated defects and frequent variations in the bronchopulmonary vasculature support the possibility that a generalized teratogenic factor may be involved (11). Pulmonary hypoplasia and severe atelectasis may present with similar clinical findings and may only be differentiated by bronchography, bronchoscopy or scanning techniques. These procedures were not performed on our patient because they were not clinically indicated. It is possible that she has severe atelectasis and not pulmonary hypoplasia, but because of her benign clinical course the diagnosis of atelectasis is unlikely.

To our knowledge this is the first reported case of Peters anomaly associated with pulmonary hypoplasia.

REFERENCES

1. Alkemade, P. P. H.: "Dysgenesis Mesodermalis of the Iris and the Cornea." Assen: Royal Van Gorcum, 1969.
2. Peters, A.: Ueber angeborene defektbildung der Descemetschen membran. Klin. M. Augenheilk. 44:27, 105, 1906.
3. Nakanishi, I. and Brown, S.: The histopathology and ultrastructure of congenital central corneal opacity (Peters anomaly). Am. J. Ophthalmol. 72:801, 1971.
4. Townsend, W. M., Font, R. L. and Zimmerman, L. E.: Congenital corneal leukomas. 2. Histopathologic findings in 19 eyes with central defect in Descemet's membrane. Am. J. Ophthalmol. 77:192, 1974.
5. Waring, G. O., Rodrigues, M. M. and Laibson, P. R.: The anterior chamber cleavage syndrome—a stepladder classification. Surv. Ophthal. 20:2, 1975.
6. Townsend, W. M.: Congenital corneal leukomas. 1. Central defect in Descemet's membrane. Am. J. Ophthalmol. 77:80, 1974.
7. Baqueiro, A. and Hein, P. A.: Familial congenital leukoma. Case report and review of the literature. Am. J. Ophthalmol. 50:810, 1960.
8. Haney, W. P. and Falls, H. F.: The occurrence of congenital keratoconus posticus circumscriptus. Am. J. Ophthalmol. 52:53, 1961.
9. Reese, A. B. and Ellsworth, R. M.: The anterior chamber cleavage syndrome. Arch. Ophthalmol. 75:307, 1966.
10. Avery, M. E. and Fletcher, B. D.: "The Lung and Its Disorders in the Newborn Infant." Philadelphia: W. B. Saunders Co., 1974.
11. Kindig, E. L., Jr.: "Disorders of the Respiratory Tract in Children," "Pulmonary Disorders." Philadelphia: W. B. Sauders, 1972, vol. 1.

Peripheral Pulmonary Cystic Disease in Sibs

Elizabeth J. Ives, FRCP (C), Maria Darja, MD
and Susana Geist, MD

The literature on pulmonary cystic disease is confused with almost as many classifications as there are authors. In general, however, cysts have been described according to their location, whether they were single or multiple and whether they were thought to be congenital or acquired. MacRae first drew attention to possible genetic factors with his 1947 report of 4 similarly affected brothers (1). Since that time there have been very few additional reports of familial instances of pulmonary cysts and it is certainly a rare cause of death in infancy. On this account we feel it is useful to report a recently encountered family in which 2 consecutive sibs have died in infancy with generalized peripheral pulmonary cystic disease.

CASE REPORTS

The main aspects of the histories of the 2 affected sibs are summarized in Table 1. At the time of birth of *Case 2*, the mother, age 24 years, and the father, age 33 years, were both healthy. The pedigree (Fig. 1) shows the lack of demonstrable consanguinity and also that the mother had previously had 2 normal children by men other than her husband.

As can be seen in Table 1, pregnancy and birth were straightforward for each affected infant. The first developed a right pneumothorax at 5 hours of age and despite satisfactory reexpansion of the lung following treatment she continued to have a rapid respiratory rate. She was referred for investigation at 2 weeks of age. All additional investigations were normal but the baby continued to thrive poorly, eventually became cyanosed and finally died with unexplained tachypnea at 3½ months. The second infant had respiratory distress from birth, also developed a right pneumothorax and despite fairly good reexpansion of that lung, had constant cyanosis and deteriorated until death at 20 hours. Apart from the evidence of pneumothorax, radiology in both these infants was consistently normal. The second baby had an accessory thumb on one hand but there were no other external anomalies in either case.

Birth Defects: Original Article Series, Volume XII, Number 5, pages 187–191
© 1976 The National Foundation

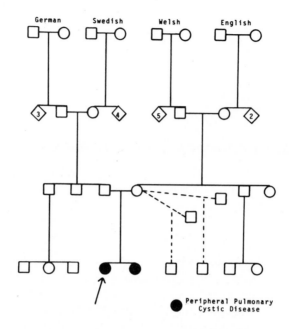

Fig. 1. Simplified pedigree of 2 affected sibs.

TABLE 1. Sibs With Pulmonary Cystic Disease

	Case 1	Case 2
Birth	November 1973	January 1975
Pregnancy	Normal	Normal
Gestation	42 wk	38 wk
Birthweight	3,380 gm	3,305 gm
Apgar	9 and 10	5 and 6
External Anomalies	None	Bifid right thumb
Postnatal Course	Right pneumothorax at 5 hr	Respiratory distress
	Persistent tachypnea	Right pneumothorax at 1½ hr
	Full investigation – no cause	Continued cyanosis and deterioration
	Some improvement – discharged	Death at 20 hr
	Thrived poorly – readmitted at 10 wk	
	Persistent tachypnea and cyanosis	
	Death at 15 wk	

Pathology

In the first case there was no disturbance to the gross anatomy of the lungs which, however, were hemorrhagic and had a finely pebbled surface. Microscopically, in the subpleural and interlobar fissure areas of both lungs but most marked on the right, there were numerous cystic spaces (Fig. 2). These cysts occasionally appeared to connect with each other but not with the airways (Fig. 3). The cysts were generally lined by cuboidal nonciliated cells (Fig. 4) and there were no adenomatous elements, thus placing them in the bronchogenic cyst category. Amorphous granular material was present in many cysts, but while there was some hemorrhage in the airways and peripheral cysts, there was only minimal evidence of infection.

In the second case the gross anatomy of the lungs was again normal. The histology in this case was much more difficult to interpret due to the poor expansion, congestion and distension of alveolar ducts with squamous and amniotic fluid debris. However, again in the subpleural and interlobar fissure regions, there were multiple areas of irregular dilations separated by branching septa (Fig. 5). It seems likely that had the infant lived longer the appearances would have become similar to those in the first case. The second case also had a large atrial septal defect; otherwise, neither child had any significant pathology elsewhere.

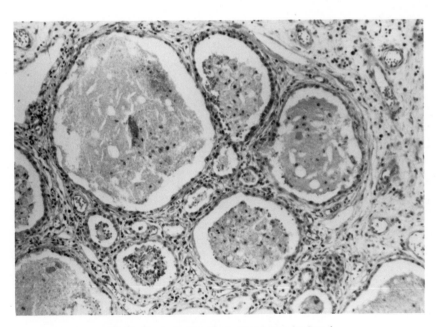

Fig. 2. General view of subpleural cysts in *Case 1.*

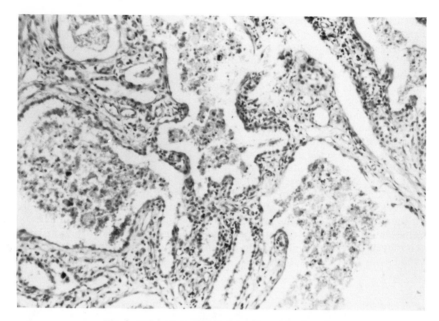

Fig. 3. Interconnecting peripheral cysts in *Case 1*.

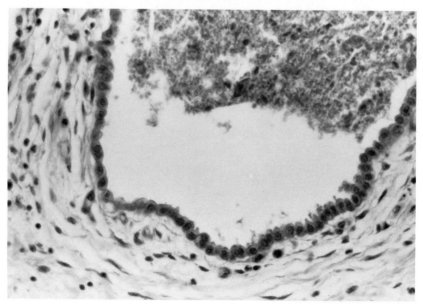

Fig. 4. High power view of cuboidal nonciliated cells lining cyst in *Case 1*.

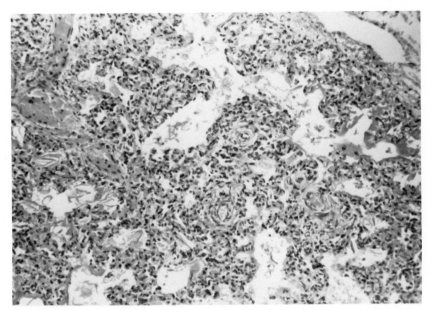

Fig. 5. Peripheral area of lung in *Case 2* showing irregularly dilated spaces, congestion and amniotic debris.

DISCUSSION

The generalized peripheral pulmonary cystic disease in these sibs seems quite different both in clinical course and histology from other reported familial cases (2). It is also different from the cystic disease which has a high incidence in Israel among Oriental Jews (3). The pathologic picture is also quite different from pulmonary lymphatic cystic disease which can give a similar clinical picture. Neither infant had any evidence of conditions such as Marfan syndrome or tuberous sclerosis in which pulmonary cysts have been described.

In the infants reported here, most of the lung appears to be normally developed implying that the origin of the cysts is probably a relatively late occurring abnormality of fetal bronchial growth. Development of respiratory elements normally continues well after birth and it seems possible that the cysts continue to progress postnatally, accounting for the difference in postmortem appearances in these 2 cases.

While it cannot be proved, we have considered autosomal recessive inheritance as most likely for the purpose of genetic counseling for the parents of these affected infants.

REFERENCES

1. MacRae, D. F.: Congenital cystic disease of the lung. Canad. Med. Ass. J. 57:545, 1947.
2. Warkany, J.: "Congenital Malformations." Chicago: Year Book Medical Publishers, 1971, p. 611.
3. Baum, G. L. et al: Cystic disease of the lung. Am. J. Med. 40:578, 1966.

A Syndrome of Ankylosis, Facial Anomalies and Pulmonary Hypoplasia Secondary to Fetal Neuromuscular Dysfunction*

Alan D. Mease, Cpt, MC, USA, Gentry W. Yeatman, Maj, MC, USA, Gary Pettett, Maj, MC, USA and Gerald B. Merenstein, Ltc, MC, USA

Arthrogryposis multiplex congenita (AMC) is a symptom complex of non-progressive, multiple congenital joint contractures and muscle wasting resulting from prenatal disease processes affecting the motor unit (1). Heterogeneous and often obscure neuromuscular pathology in utero have been hypothesized to cause joint deformities secondary to a lack of fetal movement (2). The experimental work of Drachman and Coulombre (3) who produced an arthrogryposis-like syndrome in chicks by embryonic exposure to neuromuscular blocking agents adds support to this hypothesis. Within the symptom complex of AMC the majority of patients have evidence of a primary neurogenic process (1, 4) affecting the anterior horn cells, whereas primary myopathy (1) and radiculopathy (5) have been implicated less often.

A familial pattern of malformation including ankylosis, facial anomalies and pulmonary hypoplasia has been identified within the symptom complex of AMC (6, 7). An additional family exhibiting recessive transmission of a similar malformation syndrome is described here, with studies in one case suggestive of a congenital myopathy.

CASE REPORTS

Case 1

The patient, a 2810 gm black male, was born uneventfully at 39 weeks' gesta-

*The opinions or assertions contained herein are the private views of the authors and are not to be construed as official or as reflecting the views of the Department of the Army or the Department of Defense.

tion to a 25-year-old gravida 3, para 2, ab 1 black female following an uncomplicated pregnancy. There was no evidence of oligohydramnios, although the amniotic fluid was meconium stained, and no meconium was found beneath the vocal cords. Apgars were 1 at 1 minute, 3 at 5 minutes and 7 at 10 minutes. Birthweight was at the 25th percentile, length 52.5 cm (90th%) and head circumference 35.5 cm (90th%). Physical examination revealed hypertelorism, low-set ears with occipital rotation and a cleft palate. Both lower limbs had decreased muscle mass, flexion contractures of the hips and bilateral clubfeet. The upper limbs had flexion contractures at the elbows and bilateral camptodactyly. The testes were undescended bilaterally. A chest radiograph revealed hypoexpansion of the right hemithorax and a mild thoracic kyphosis (Fig. 1).

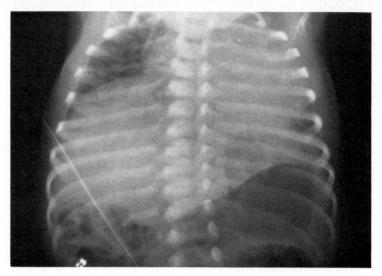

Fig. 1. *Case 1*, AP chest radiograph. Note elevation of right hemidiaphragm.

The infant's hospital course was complicated by supplemental oxygen requirement for the first 48 hours of life, at which time a respiratory arrest necessitated ventilatory assistance. A tracheostomy was performed at 13 days of age because of continued need for respirator support. At 20 days of age the infant developed a right tension pneumothorax, and a chest tube was inserted. The infant died at 24 days of age of probable sepsis, despite several courses of antibiotics throughout his hospital course.

At autopsy the lungs were congested and firm. A microscopic examination of the lungs revealed inflammatory infiltrates bilaterally with poorly developed alveoli in the right lung. The diaphragms were intact bilaterally. The heart was normal except for a patent ductus arteriosus. The kidneys were grossly and histologically normal. Pathologic studies of peripheral muscle and anterior horn cells were not done.

Chromosomal analysis of this infant's peripheral blood revealed a 46,XY pattern without structural anomalies.

Case 2

Case 2, a 2215 gm black male infant, was born 12 months after *Case 1*, following an uncomplicated pregnancy of 38 weeks. Labor and delivery were complicated by frank breech presentation. The infant was delivered vaginally with Apgars of 3 at 1 minute and 7 at 5 minutes. There was no evidence of oligohydramnios. Retrospectively, the mother noted a relatively decreased amount of fetal movement during the pregnancies with *Case 1 and 2* compared to that of an older normal female sib.

The birthweight was less than the 10th percentile, length 51 cm (97th%) and head circumference 35.5 cm (90th%). On physical examination the infant had hypertelorism, low-set ears with occipital rotation, ptosis of the left eye, a midline port-wine hemangioma of the forehead and a high-arched palate, without a cleft. There was markedly decreased muscle mass of all limbs with bilateral ankyloses of the knees and bilateral flexion contractures of the hips. The elbows had prominent dimples with webbing and flexion contractures bilaterally. An examination of the hands revealed mild camptodactyly (Fig. 2). The testes were un-

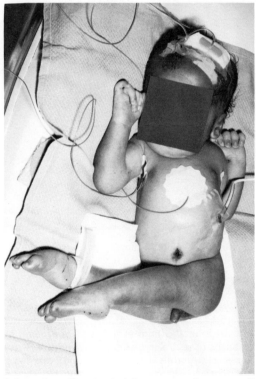

Fig. 2. *Case 2*, 3 days of age. Note marked decrease muscle mass on all limbs, bilateral anklyoses of the knees and bilateral flexion contractures of hips and elbows.

descended bilaterally. Chest radiographs revealed a small bell-shaped thoracic cage with moderate thoracic kyphosis.

The neonatal course was complicated by supplemental oxygen requirement for the first 13 days of life. Other problems included unproved sepsis and physiologic jaundice. The infant's cry was weak and suck was poor throughout hospitalization. In order to maintain an adequate caloric intake for growth, gavage feedings were necessary. After discharge, the infant required gavage feedings at home for 1 month.

Currently, at 12 months of age he is developmentally normal except for gross motor skills (Fig. 3) and a weight that has remained well below but parallel to the 3rd percentile. He has had no evidence of progressive weakness but has developed progression of kyphosis.

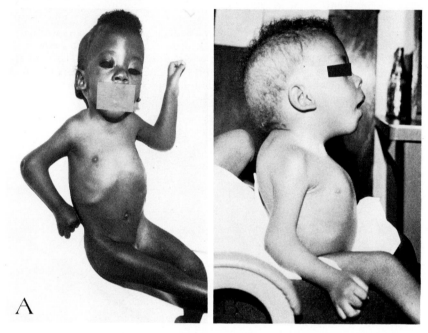

Fig. 3. *Case 2,* 12 months old. A) Note left ptosis, hypertelorism, webbing of elbows and dimple of right elbow. B) Note low-set ear and severe kyphosis.

Chromosomal analysis from peripheral blood lymphocytes revealed a 46,XY pattern without structural anomalies. Serial ECGs have been normal for age. CPK levels were 56.8 units (normal 0-50 units) at 7 days of age and 299 units (normal 25-145 units) at 3 months of age. Nerve conduction and EMG studies at 4 months of age revealed normal right peroneal nerve conduction and the presence of small polyphasic muscle potentials. The right gastrocnemius muscle was biopsied at 4 months of age.

The family pedigree reveals consanguinity, high incidence of fetal wastage, and 2 individuals with mental retardation on the maternal side (Fig. 4). The paternal pedigree is unremarkable.

The skeletal muscle biopsy was examined by the Armed Forces Institute of Pathology. A portion was routinely fixed and stained with hematoxylin and eosin, periodic acid schiff, and modified Gomori trichrome stains. Histochemical studies, including ATPase, reverse ATPase, NADH-tetrazolium reductase and phosphorylase, were performed on a quick frozen specimen. A third portion of the specimen was imbedded in epon and examined by electron microscope.

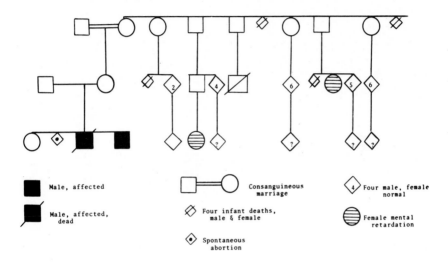

Fig. 4. Shortened pedigree of maternal kindred.

The microscopic findings show considerable variation in muscle size with most of the fibers being smaller than normal and a lesser percentage being hypertrophic (Fig. 5A). Numerous small nerve trunks were histologically normal. Small groups of fibers are seen which demonstrate striking regenerative changes consisting of enlarged vesicular nuclei with prominent nucleoli and basophilic sarcoplasm (Fig. 5B). ATPase histochemistry shows differentiation of fiber types with a striking predominance of type 1 fibers. This is confirmed by reverse ATPase and NADH-tetrazolium reductase (Fig. 6). The phosphorylase stain was unremarkable. Electron microscopic evaluation revealed nonspecific changes in areas of degeneration and regeneration without evidence of mitochondrial abnormalities.

DISCUSSION

These cases suggest an autosomal recessive pattern of inheritance of a new syndrome within the symptom complex of AMC, similar to that described by Pena and Shokeir (6) and Punnett et al (7) (Table 1). Survival of *Case 2* to 12

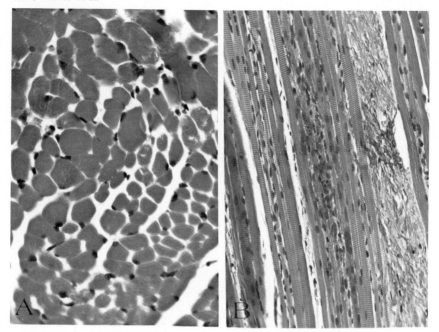

Fig. 5. *Case 2,* gastrocnemius muscle. A) Small rounded fibers with variation in diameter (H and E, × 198). B) Focus of degeneration-regeneration (H and E, ×98).

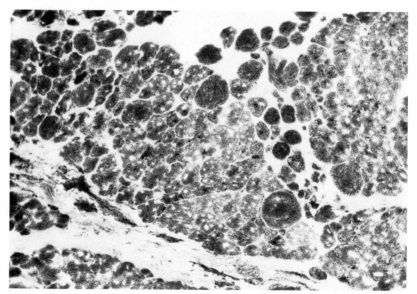

Fig. 6. *Case 2,* gastrocnemius muscle. Histochemistry with areas of larger rounded fibers and type 1 (dark and intermediate) fiber predominance (NADH-tetrazolium reductase, ×108).

TABLE 1.

Sex	Pena and Shokeir		Punnett et al		Mease et al	
	F	F	M	M	M	M
Complications of Pregnancy						
IUGR*	+	+	+	+	+	+
Perinatal death	+	+	+	+	+	−
Hydramnios	+	+	−	−	−	−
Short cord	0	0	+	+	−	−
Small placenta	0	−	+	+	−	−
Abruptio	0	+	−	−	−	−
Facies						
Low-set ears	+	+	+	+	+	+
Hypertelorism	+	+	+	+	+	+
Epicanthal folds	0	0	+	+	+	+
Depressed tip of nose	+	+	+	+	+	+
Cleft palate	−	−	−	−	+	−
Ptosis	0	0	0	0	−	+
Limb						
Muscle atrophy	+	+	+	+	+	+
Arthrogryposis	+	+	+	+	+	+
Clubfeet	+	+	+	+	+	+
Camptodactyly	+	+	+	+	+	+
Miscellaneous						
Hypoplastic lungs	+	+	+	±	+	±
Undescended testes	NA	NA	+	+	+	+

+ = Present
− = Not present
0 = Not commented on
NA = Not applicable
*Intrauterine growth retardation

months of age indicates that this syndrome is not universally fatal. Survival may be dependent upon aggressive neonatal management and extent of pulmonary pathology.

There is a striking overlap of many features of this syndrome with the non-renal features of Potter syndrome. These same features have been attributed to fetal compression secondary to oligohydramnios by Thomas and Smith (8). Oligohydramnios was not present in the 6 cases of this syndrome. Lack of fetal movement in this syndrome secondary to prenatal disease of the motor unit and secondary to fetal compression by oligohydramnios in renal agenesis may be hypothesized as an explanation for this overlap of features (Fig. 7).

In *Case 2* an attempt was made to elucidate the nature of the motor unit pathology using routine, histochemical and electron microscopic techniques. These findings were most consistent with a congenital myopathy (9, 10) but

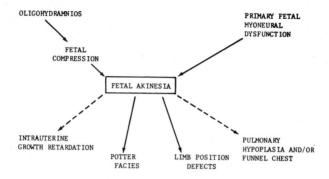

Fig. 7. Lack of fetal movement hypothesis.

anterior horn-cell disease could not be excluded. Prenatal neuromuscular dysfunction caused by familial myopathy is hypothesized as the basic defect in this syndrome.

ACKNOWLEDGMENTS

We are greatly indebted to Dr. Vernon Armbrustmacher, LTC, Chief of the Division of Neuromuscular Pathology of the Armed Forces Institute of Pathology, who performed routine, histochemical and electron microscopic studies on the muscle biopsy and assisted in interpretation of the results.

REFERENCES

1. Daentl, D. L., Berg, B. O., Layzer, R. B. and Epstein, G. J.: A new familial arthrogryposis without weakness. Neurology 24:55-60, 1974.
2. Mead, N. G., Lithgow, W. C. and Sweeney, H. J.: Arthrogryposis multiplex congenita. J. Bone Joint Surg. 40A:1285-1309, 1958.
3. Drachman, D. B. and Coulombre, A. J.: Experimental clubfoot and arthrogryposis multiplex congenita. Lancet 2:523-526, 1962.
4. Fisher, R. L., Johnstone, W. T., Fisher, W. H. and Goldkamp, O. G.: Arthrogryposis multiplex congenita: A clinical investigation. J. Pediatr. 76:255-261, 1970.
5. Pena, C. E., Miller, F., Budzelovich, G. N. and Feigin, I.: Arthrogryposis multiplex congenita. Report of two cases of a radicular type with familial incidence. Neurology 18: 926-930, 1968.
6. Pena, S. D. J. and Shokeir, M. H. K.: Syndrome of camptodactyly, multiple ankyloses, facial anomalies and pulmonary hypoplasia: A lethal condition. J. Pediatr. 85:373-375, 1974.
7. Punnett, H. H., Kistenmacher, M. L., Valdes-Dapena, M. and Ellison, R. T.: Syndrome of ankylosis, facial anomalies and pulmonary hypoplasia. J. Pediatr. 85:375–377, 1974.
8. Thomas, I. T. and Smith, D. W.: Oligohydramnios, cause of the nonrenal features of Potter's syndrome, including pulmonary hypoplasia. J. Pediatr. 84:811-814, 1974.
9. Climie, A. R. W.: Muscle biopsy: Technic and interpretation. Am. J. Clin. Pathol. 60: 753-770, 1973.
10. Brooke, M. H. and Kaiser, K. K.: The use and abuse of muscle histochemistry. Ann. N.Y. Acad. Sci. 228:121-144, 1974.

Syndrome of Camptodactyly, Multiple Ankyloses, Facial Anomalies and Pulmonary Hypoplasia — Further Delineation and Evidence for Autosomal Recessive Inheritance

Sergio D. J. Pena, MD* and Mohamed H. K. Shokeir, MD, PhD†

Recently we described 2 sisters who displayed an identical picture of severe camptodactyly, clubfeet, knee and hip ankyloses, characteristic facies, pulmonary hypoplasia and perinatal death (1). Our suggestion that this pattern of malformations constituted a new lethal syndrome probably inherited as autosomal recessive trait received prompt support from Punnett et al (2) who reported 2 sporadic cases. However, possible teratogenic causation was put forward as an alternative etiologic possibility.

We report 3 other patients afflicted with the disorder, bringing the total number of cases to 7. These observations permit further delineation of the syndrome and provide strong evidence of autosomal recessive inheritance rather than a teratogenic causation.

CASE REPORTS

Case 1

This 1075 gm male infant was born by precipitate delivery after spontaneous premature rupture of the membranes; the mother was unaware of her last menstrual period. Despite resuscitative measures the child expired at 1 hour of age. Crown-heel length was 38 cm; crown-rump length was 26 cm; head circumference was 27 cm. Physical examination disclosed hypertelorism, low-set malformed ears, a depressed tip of the nose, high-arched palate and micrognathia. There was ulnar deviation of the hands and severe camptodactyly, with the thumb and the 5th digit overlapping the other fingers. The hips could not be

*Supported by a Post-Doctoral Fellowship of the MRC of Canada
†Queen Elizabeth II Scientist

abducted more than 45°; knees were held in extension; and feet displayed talipes equinovarus with metatarsus varus. Testes were undescended. Autopsy revealed hypoplastic lungs with a combined weight of 14.6 gm (expected for weight 29 gm^3), and an atrial septal defect; kidneys were normal. The placenta (290 gm) showed marked loss of the cotyledon pattern and moderate degenerative changes of the villi. Chromosome studies showed a 46, XY karyotype with normal G-banding pattern.

Case 2

This 1260 gm female infant (sister of *Case 1*) was spontaneously delivered at 37 weeks of gestation after a pregnancy complicated by hydramnios and mild toxemia. Infant expired at 1 hour of life in severe respiratory distress. Crown-heel length was 38 cm; crown-rump length 25.5 cm; head circumference 27.5 cm; intercanthal distance 2.1 cm (expected for age 1.6 cm). Physical examination disclosed low-set malformed ears, hypertelorism, a flat face with depressed nose tip, high-arched palate, and micrognathia (Figs. 1 and 2). Upper limbs were held straight along body with marked ankylosis of the elbow; hands exhibited camptodactyly and clinodactyly, hypoplastic dermal ridges and virtually absent palmar creases. There was mild thoracic kyphosis. The lower limbs were ankylosed in extension and the feet had a rocker-bottom deformity. At autopsy the lungs showed a combined weight of 28 gm (expected for weight 29 gm; expected for age 37 gm^3); heart and kidneys were normal. A subarachnoid hemorrhage was present. The placenta (270 gm) showed moderate loss of cotyledon pattern but was normal at microscopy. Karyotype was 46,XX with a normal G-banding pattern.

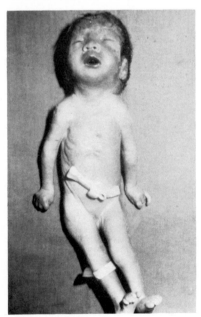

Fig. 1. *Case 2.*

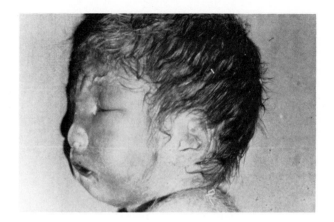

Fig. 2. *Case 2.*

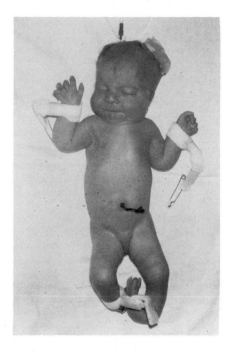

Fig. 3. *Case 3.*

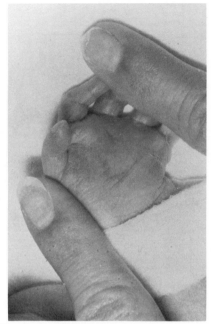

Fig. 4. *Case 3.* Note very faint palmar creases.

Case 3

This 2050 gm female infant was born at 38 weeks of gestation after a pregnancy complicated by hydramnios. Delivery was by Kjellands forceps for a transverse lie. Placenta was small. Crown-heel length was 41 cm; crown-rump length 30 cm; head circumference 30 cm. Physical examination revealed low-set malformed ears, hypertelorism, depressed tip of the nose, a normal palate and micrognathia (Fig. 3). Hands displayed camptodactyly; the palmar creases were barely visible (Fig. 4); and the dermal ridges were hypoplastic. Abduction of the hips was limited and the knees were ankylosed in extension. There was severe clubfeet. The child presented cyanosis and congestive heart failure and expired at 3 days with necrotising enterocolitis. Autopsy disclosed complete transposition of the great vessel with pulmonic stenosis, incomplete malrotation of the small intestine and a subarachnoid hemorrhage. Lungs were severely congested and their combined weight was 35 gm (expected for weight 44 gm^3). Karyotype was 46,XX with normal G-banding pattern.

Family Histories

Cases 1 and 2 were the product of an incestuous uncle-niece relationship. The gravida 2, para 2, ab 0 mother was 16 and 17 years old respectively at the birth of the 2 infants. She suffered from mild idiopathic epilepsy and intermittently used Diphenylhydantoin during the 2 pregnancies. There was no family history of birth defects.

Case 3 was born to a 23-year-old gravida 2, para 0, ab 1 female. An abortion, 3 years earlier, was induced. There was no drug ingestion or x-ray exposure in the 2nd pregnancy. Parents are unrelated and there is no family history of similar abnormaiities.

DISCUSSION

The initial delineation of a new syndrome is likely to be extensively modified as further cases appear; one reason for this trend is an ascertainment bias inasmuch as the patients who initially brought the malformation pattern to light are likely to be among the more severely afflicted. Accordingly, the 2 initial cases of this syndrome had an extreme degree of pulmonary hypoplasia (1). However, later cases have shown less pronounced degrees of lung involvement. The definition of what constitutes pulmonary hypoplasia becomes critical in deciding on whether or not the anomaly actually exists. If a criterion based on lung mass is adopted, reference to standard tables of weights of organs of fetuses and infants such as the one by Schulz et al (3) may be contemplated. However, we found that standard deviations are generally so large as to limit their usefulness. Potter (4) defines pulmonary hypoplasia on histologic as well as lung mass criteria, and uses the rather convenient lung weight/body weight ratio (hypoplasia is considered to exist if the ratio is less than 0.013). We decided to use this ratio, waiving the more subjective histologic requirements. By doing so, Case 2 of Punnett et al (2) and *Cases 2 and 3* of the present report do not have pulmonary hypoplasia (see Table 1). Otherwise, these cases do fit the delineated syndrome, and the sib relationship of *Cases 1 and 2* of this report further emphasizes that lung hypoplasia is a variable manifestation of the disorder. Another feature of the affliction, namely hypoplasia of hand ridges and creases, was first pointed out by Punnett et al (2) and found in 2 of the current patients and in Case 2 of our original report. Also cryptorchidism, placental anomalies and hydramnios have now been confirmed as components of the syndrome.

An outline of the syndrome as it currently stands is given in Table 1 and Figure 5.

Finally, the close parental consanguinity of *Cases 1 and 2* of the present report, combined with the consistently normal parental phenotype, the affliction of both sexes and the absence of any identifiable common environmental agent, provide evidence strongly suggestive of autosomal recessive inheritance as originally proposed, rather than a teratogenic causation.

ACKNOWLEDGMENT

We are grateful to Dr. D. Grewar for his cooperation in this study.

TABLE 1. Comparison of the Seven Patients

Abnormalities	Pena and Shokeir (1)		Punnett et al (2)		Present report			Total
	Case 1	Case 2	Case 1	Case 2	Case 1	Case 2	Case 3	
	female	female	male	male	male	female	female	female:male = 4:3
Complications of pregnancy								
Intrauterine growth retardation	+	+	+	+	+	+	+	7/7
Perinatal death	+	+	−	+	+	+	+	6/7
Placental abnormalities or								
small placenta	?	+	+	+	?	+	+	7/7
Hydramnios	+	+	−	−	?	+	+	4/6
Facies								
Hypertelorism	+	+	+	+	+	+	+	7/7
Low-set malformed ears	+	+	+	+	+	+	+	7/7
Depressed tip of the nose	+	+	+	+	+	+	+	7/7
Micrognathia	+	+	+	+	+	+	+	7/7
Extremities								
Camptodactyly	+	+	+	+	+	+	+	7/7
Hypoplastic ridges and creases	?	+	+	+	?	+	+	5/5
Ulnar deviation of hands	+	+	+	−	+	−	−	4/7
Ankyloses	+	+	+	+	+	+	+	7/7
Clubfeet	+	+	+	+	+	−	+	6/7
Other Anomalies								
Hypoplastic lungs	+ (0.003*)	+ (0.003)	+ (0.012)	− (0.016)	+ (0.013)	− (0.022)	− (0.017)	4/7
Cryptorchidism	NA	NA	+	+	+	NA	NA	3/3

*Lung weight/body weight ratio
NA = Not Applicable

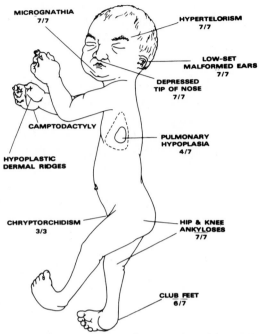

Fig. 5. Main dysmorphic features of the syndrome. [Adapted from *J. Pediatr. 85:*374, 1974 (1)].

COMMENTS

Peter A. Bowen: Two black female sibs with many of the features described in this paper were reported in the Bulletin of the Johns Hopkins Hospital 114: 402–414, 1964. In addition to prominent forehead, flat nose, camptodactyly and flexion contractures of the wrists, elbows and knees, both sibs had ventral abdominal hernias, agenesis of the corpus callosum and congenital glaucoma. One had agenesis of segments or lobes of the lungs, platybasia, partial arrhinencephaly, and anomalous pulmonary venous drainage; the other had ureterovesical reflux and dilatation of the ureters and renal pelves.

Sergio D. J. Pena: In the cases reported by Dr. Bowen there is ectodermal involvement as shown by CNS malformations, in contrast to our syndrome in which it seems that only mesodermal derivatives are affected. The infants reported by Bowen et al, displayed, besides CNS deformities, involvement of the eyes and anterior abdominal wall — none of these have been noticed in the reported cases of the disorder described by us. Their patients lacked the characteristic hydramnios, placental anomalies and intrauterine growth retardation seen in our syndrome. Unfortunately, the cases of Bowen et al were reported before the advent of chromosome banding techniques and a karyotypic abnormality cannot be ruled out.

In reviewing our medical records on pulmonary hypoplasia, we came across a case with pulmonary hypoplasia, renal cysts, multiple ankyloses and arrhinencephaly. We did not feel that this patient suffered from the syndrome described by us; the clinical picture nonetheless seems similar to the infants reported by Bowen et al. Unfortunately, photographs were not available.

REFERENCES

1. Pena, S. D. J. and Shokeir, M. H. K.: Syndrome of camptodactyly, multiple ankyloses, facial anomalies, and pulmonary hypoplasia: A lethal condition. J. Pediatr. 85:373, 1974.
2. Punnett, H. H., Kistenmacher, M. L., Valdes-Dapena, M. and Ellison, R. T.: Syndrome of ankylosis, facial anomalies, and pulmonary hypoplasia. J. Pediatr. 85:375, 1974.
3. Schulz, D. M., Giordano, D. A. and Schulz, D. H.: Weights of organs of fetuses and infants. Arch. Pathol. 74:244, 1962.
4. Potter, E. L.: "Pathology of the Fetus and Infant," 2nd Ed. Chicago: Yearbook Medical Publishers, 1961.

Accelerated Skeletal Maturation Syndrome With Pulmonary Hypertension

Jane C. S. Perrin, MD, Edgardo Arcinue, MD, William H. Hoffman, MD, Harold Chen, MD and Joseph O. Reed, MD

Since 1971, 4 cases have been reported of a rare syndrome including advanced bone age, peculiar facies, failure to thrive and respiratory problems (1–3). We have had such an infant under our care from birth to 3 months who has, in addition, pulmonary hypertension.

CASE REPORT

The patient is a black male born in December 1974 to a 20-year-old gravida 4, para 3, ab 1 mother and 30-year-old healthy father. The parents are unrelated and 2 female sibs are normal. A 16-year-old maternal uncle is retarded, and a maternal aunt who died of leukemia at age 22 had 2 stillborn sons plus a normal one.

The mother was drug dependent throughout pregnancy – on heroin the first 6 months' gestation and heroin plus methadone 15–20 mg/day the third trimester. VDRL titer prior to delivery was 1:32, and urine was positive for quinine and heroin up to one month prior to delivery. Following rupture of membranes and 16 hours of labor at term, fetal distress developed and the infant was delivered by C section. The 1-minute and 5-minute Apgar scores were 6 and 8. Birthweight was 3100 gm (40%), length 53 cm (75%), head circumference 38 cm (97%) and maturity score 39 weeks' gestation.

Unusual physical features (Figs. 1–3) included frontal prominence and sagittal ridging of the skull, flat nasal bridge, prominent eyes with megacornea (12 mm), narrow thorax, stridor, grade 2/6 systolic heart murmur, stretched phallic length of 5 cm (> 2 SD) (4) with an ectopic eccentric meatus, hirsute limbs with long bulbous fingers and toes, spoon nails, hypoplastic palmar creases and dermal ridge patterns, hypotonia and poor suck.

Birth Defects: Original Article Series, Volume XII, Number 5, pages 209–217
© 1976 The National Foundation

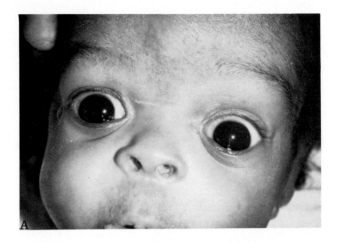

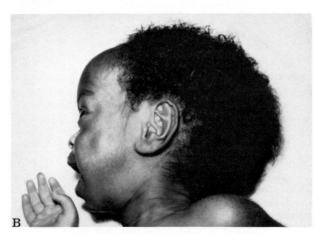

Fig. 1. A) Face showing megacornea. B) Profile showing large head.

On roentgenologic skeletal survey at birth (Fig. 4), bone age was 21 months by count of left hemiskeleton epiphyses (5), and left hand and wrist bone age was 15–18 months by Atlas standards (6). Radial length was 63 mm (average for newborn 58); there were no syphilitic bone changes. The proximal and middle phalanges of the hand were thickened and elongated. The thoracic and lumbar vertebral end-plates were slightly concave, and the skull showed frontal bossing and premature closure of the posterior segment of the sagittal suture. Echoencephalogram indicated a 10 mm wide 3rd ventricle with no midline shift. The IVP was normal.

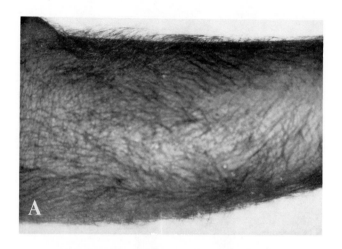

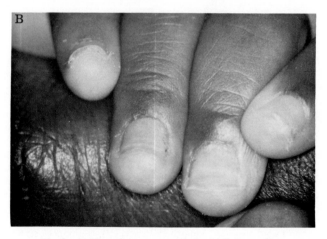

Fig. 2. A) Hirsutism on upper limb. B) Spoon nails.

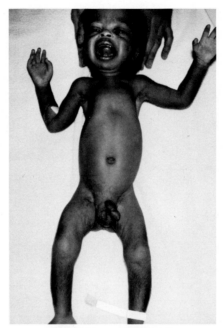

Fig. 3. Large rotated penis.

Normal or negative laboratory values the first month of life included urine amino acid chromatography, chromosome karyotype 46,XY; absent urine quinine and heroin; serum total protein 5.9 gm%, albumin 3.9, globulin 2.0, normal electrophoretic pattern; total lipid 250 mg%; glucose 120 mg%, BUN 15 mg%, serum electrolytes normal range, calcium 9.7 mg%, alkaline phosphatase 32.5 mU/ml, SGOT 31 IU/L; urine blood and cerebral spinal fluid cultures no growth; cerebral spinal fluid protein 93 mg%. Urine pregnanetriol was < 0.02 mg/24 hr (normal < 0.4 mg), 17-KS 0.9 mg/24 hr (< 2 mg/24 hr in first 3 weeks of life), 17 hydroxysteroids 0.5 and 0.3 mg/24 hr (3 ± 1 mg/M^2/24 hr). Plasma HGH was 17.5 ng (normal for newborn), TSH 5.5 and 3.8 IU/ml (< 7.2 IU/ml); serum thyroxine 7.1 ug% (2.4–7.4 ug%). Serology ART was positive 1:4 (the infant was treated with methicillin).

By age 3 months the patient, taking oral feedings amounting to 150 cal/kg/day for a month, had failed to grow. Length (53 cm), U/L segments (30.5/22.5 cm) and head circumference (38 cm) were the same as at birth, and weight was only 380 gm above birthweight. Peripheral BP was normal. In spite of this growth failure, bone age had advanced to 36 months by epiphyseal center count and to 30 months by Atlas hand and wrist standards (Fig. 5). Radial length had increased only 1 mm to 64 mm (average 66). He had passed no developmental milestones and remained hypotonic with brisk reflexes but no spasticity. EEG was normal. Ophthalmalogic exam at that time showed lid retraction, megacornea but no glaucoma, normal fundi, but no following of light or motion; globes were unequal in AP diameter on echogram.

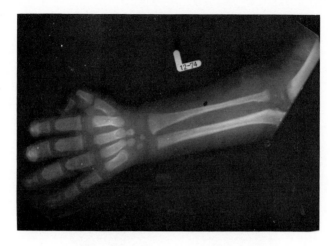

Fig. 4. Radiograph of left hand and wrist at birth: bone age 15–18 months; thickened and elongated phalanges.

   The pulmonary problem had 3 facets: pulmonary artery hypertension, upper airway obstruction, and an abnormal alveolar-arterial oxygen tension gradient ($A-aDO_2$). The grade 2/6 systolic murmur along the left sternal border was associated with severe right ventricular hypertrophy on ECG and a markedly dilated pulmonary artery on x-ray examination (Fig. 6). Pulse was not bounding but the blood pressure showed a slight pulse pressure difference (120/70 mm Hg). Cardiac catheterization showed a small 1.8 mm patent ductus arteriosus, no right-to-left shunt and a small left-to-right shunt. There was a marked fluctuation of the venae cavae and atrial pressures secondary to breathing difficulties and marked swings of intrathoracic pressures. The main pulmonary artery pressure was 90/50 (mean 63), compared to the aortic pressure of 120/70 (mean 90), showing pulmonary hypertension. Oxygen saturations were highly variable because of breathing instability and thus did not allow accurate pulmonary flow and resistance measurements. Overall evaluation of the catheterization data gave the impression that the pulmonary artery hypertension was not secondary to the left-to-right shunt through the ductus arteriosus, but rather to the upper airway obstruction.
   Stridor was a prominent feature throughout the clinical course. Direct laryngoscopy showed a normal glottis and epiglottis; the posterior pharynx was narrow, including the tonsillar fauces. The tongue was big in relation to the pharynx, and stridor was most prominent when the patient was supine, with the tongue nearly occluding the oropharynx because it was sucked in as the patient

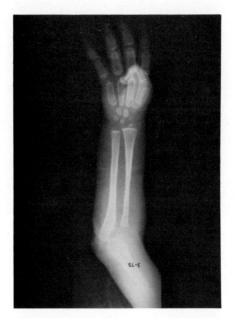

Fig. 5. Left hand at 3 months: bone age 30 months.

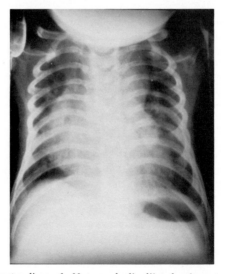

Fig. 6. Chest radiograph. Note markedly dilated pulmonary artery.

fought to breathe. In addition a 2.5 mm nasal tube passed with difficulty, although choanal stenosis was not confirmed on nasogram.

At age 2 months an attempt to relieve airway obstruction by suturing the tongue in the midtongue position to the lip was unsuccessful as the tongue continued to fall back into the pharynx. Tracheostomy was therefore performed. Prior to tracheostomy, arterial CO_2 levels ($PaCO_2$) were high in the range of 38–62 mm Hg. Respiratory acidosis with a pH of 7.15 was seen occasionally but most of the time compensated pH was the rule. After tracheostomy the highest $PaCO_2$ was 45 (range 23–45). Seventeen days after tracheostomy when the lungs appeared relatively clear by x-ray examination and upper airway obstruction was absent, the arterial blood gases at 21% O_2 showed pH 7.32, pCO_2 32, pO_2 53; at 100% O_2 they were pH 7.48, pCO_2 25, pO_2 333. The calculated A-aDO_2 was 63 in room air (normal 15) and 349 in 100% oxygen (normal 70). This suggested a large intrapulmonary venous admixture, and since arterial blood was drawn from the right brachial artery, a right-to-left shunt in the patent ductus arteriosus was ruled out.

AUTOPSY REPORT

The infant died at 6 months. At autopsy he was poorly nourished with a length of 62 cm, CR length 43 cm, head circumference 41 cm and a narrow chest of 30 cm. Significant findings were atelectasis and multifocal aspiration pneumonitis. The enlarged heart had a patent but valvular competent foramen ovale and probe patent ductus arteriosus; there was right ventricular hypertrophy and dilated main and right and left pulmonary arteries which compressed the mainstem bronchi. Many medium and small pulmonary arteries had medial hypertrophy.

The tongue was large and tracheostomy stoma margins healed. There was no thyroid, parathyroid or adrenal gland abnormality.

There was marked advanced ossification of epiphyseal centers in the humeral head and distal femur upon examination. Microscopically, the zone of provisional calcification appeared foreshortened. The cartilaginous columns were well aligned but short in the epiphysis and islands of cartilage persisted in the osteoid trabeculas.

The skull showed premature closure of the sagittal suture and a long narrow foramen magnum. Brainweight was 713 gm (N:660), convolutions were normally developed, lateral ventricles moderately dilated, white matter thinned, aqueduct patent and pituitary normal. Microscopic exam showed only mild astrocytic gliosis and individual neuronal necrosis. The ventricular enlargement appeared compensatory to reduced white matter.

DISCUSSION

Exchange of gases in alveoli is by pressure gradients (7). A perfect system would be one where no gradients exist between alveolar gas concentration and final arterial blood concentration. Allowing for the normal physiologic shunt of 2–5%, there is a normal A-aDO_2 of 15 mm Hg at breathing room air and 70 mm

Hg at 100% O_2. In this patient, the A-aDO_2 was observed to be larger than normal and the wide spread in values in room air and 100% oxygen suggested underperfusion because of intrapulmonary shunting from pulmonary hypertension and microatelectasis.

The patient's mother took heroin throughout pregnancy. Heroin (diacetylmorphine) crosses the placenta (8), but is an unlikely cause of the respiratory disease in this infant. Premature infants of addicted mothers have an unexpectedly low incidence of respiratory distress syndrome, and some experimental evidence indicates heroin induces surfactant synthesis (9).

Quinine (heroin filler) has been implicated as a human teratogenic agent, possibly causing congenital deafness but not linked to any definite pattern of congenital malformation (10). Congenital defects which have been observed in infants following attempted abortion with quinine include tracheoesophageal fistula, ambiguous genitalia, congenital heart disease and reduction malformations of the limbs. None of these structural abnormalities except a small patent ductus were present in our patient.

Heroin is not a known human teratogenic agent. Harper et al (11) studied 51 pregnant addicts on heroin, methadone or both, who delivered liveborn infants. Discounting one with polydactyly (high prevalence in black population), 2 infants had congenital malformations: one with diaphragmatic hernia, the other with thoracic hemivertebra. This incidence of 4% is not significantly greater than general population figures.

Because previously reported cases have not been infants of maternal addicts, drug dependence and the occurrence of the syndrome complex in our case was probably coincidental. Although 4 out of 5 reported infants are males, there has been no recurrence reported in a family to suggest X-linked or other hereditary patterns.

Accelerated skeletal maturation may be secondary to endocrine, metabolic or intrinsic osseous defects. Although the 3 males reported previously have not had penile enlargement, our endocrine data failed to suggest either an 11β-hydroxylase or 21-hydroxylase deficiency. This is in keeping with the observation that congenital adrenal hyperplasia rarely presents in the newborn period with such accelerated skeletal maturation. Necropsy on our patient showed normal endocrine glands and minor intrinsic bone abnormalities. The brain atrophy did not account for immediate cause of either advanced epiphyseal calcification or failure to thrive.

ACKNOWLEDGMENT

Cardiac catheterization and angiograms were performed by Dr. Zia Farooki.

REFERENCES

1. Marshall, R. E., Graham, C. B., Scott, C. R. and Smith, D. W.: Syndrome of accelerated skeletal maturation and relative failure to thrive: A newly recognized clinical growth disorder. J. Pediatr. 78:95–101, 1971.
2. Tipton, R. E., Wilroy, R. S., Jr. and Summitt, R. L.: Accelerated skeletal maturation in infancy syndrome: Report of a third case. J. Pediatr. 83:829–832, 1973.
3. Visveshiwara, N., Rudolph, N. and Dagutsky, D.: Syndrome of accelerated skeletal maturation in infancy, peculiar facies, and multiple congenital anomalies. J. Pediatr. 84: 553–556, 1974.
4. Feldman, K. W. and Smith, D. W.: Fetal phallic growth and penile standards for newborn male infants. J. Pediatr. 86:395–398, 1975.
5. Sontag, L. W., Snell, D. and Anderson, M.: Rate of appearance of ossification centers from birth to age 5 years. Am. J. Dis. Child. 58:949, 1939.
6. Greulich, W. W. and Pyle, S. I.: "Radiographic Atlas of Skeletal Development of the Hand and Wrist." Palo Alto: Stanford University Press, 1959.
7. Nunn, J. F. "Applied Respiratory Physiology." New York: Appleton-Century-Crofts, 1969.
8. Glass, L., Rajegowda, B. K., Mukherjee, T. K. et al: Effect of heroin on corticosteroid production in pregnant addicts and theri fetuses. Am. J. Obstet. Gynecol. 117:416–418, 1973.
9. Taursch, H. W., Jr., Carson, S. H., Wang, N. S. and Avery, M. E.: Heroin induction of lung maturation and growth retardation in fetal rabbits. J. Pediatr. 82:869–875, 1973.
10. Warkany, J.: "Congenital Malformations." Chicago: Year Book Medical Publishers, Inc., 1971.
11. Harper, R. G., Solisk, G. I., Purow, H. M. et al: The effect of a methadone treatment program upon pregnant heroin addicts and their newborn infants. Pediatrics 54:100–305, 1974.

A Syndrome Manifested by Brittle Hair With Morphologic and Biochemical Abnormalities, Developmental Delay and Normal Stature*

Amir I. Arbisser, MD, Charles I. Scott, Jr., MD, R. Rodney Howell, MD, Poen S. Ong, PhD and Hollace L. Cox, Jr., PhD

We are reporting 2 sibs with a syndrome manifested by abnormal hair, mental deficit, normal stature and no other apparent structural or ectodermal defects. We are also reporting a unique application of a new x-ray energy dispersive fluorescent spectrometer as an analytic technique for human hair trace element analysis. Familial investigation suggests a recessive mode of inheritance.

CASE REPORTS

Case 1

The proband was a 5-month-old girl born 3/14/75, the product of an uncomplicated 40-week gestation. Labor was spontaneous and delivery was vertex. The mother's age was 28 and the father's 31. Birthweight was 3374 gm and length 48.3 cm. The patient followed an uncomplicated neonatal course and was discharged from the hospital at 1 day of age. No abnormalities were noted at birth other than brittle hair similar to her oldest brother's.

The patient was first seen in the Pediatric Outpatient Clinic because of diarrhea which subsequently cleared. At that time dry brittle hair and occipital alopecia were noted. Family history revealed a normal male sib and parents, as well as a similarly affected older brother (*Case 2*). Though the family history is negative for consanguinity, the family had recently moved from a small remote village in northern Mexico in which there are reportedly 11 similarly affected children in other families.

At age 5 months, physical examination showed a head circumference of 42

---

*This research was supported in part by National Institute of General Medical Sciences grants GM 19513 and GM 00143, and grant G 545 from the Robert A. Welch Foundation.

Birth Defects: Original Article Series, Volume XII, Number 5, pages 219–228

cm, length 65 cm, span 62 cm and weight 7527 gm. In general the patient was a large infant with abnormal hair (Fig. 1). The patient had broad symmetric facies. The anterior fontanel measured 2 × 3 cm; the posterior fontanel was closed. There was very little hair on the head and that present was short, dry, fragile, and broke easily; it was essentially absent in the areas of contact with a pillow which were covered with multiple hair fragments (Fig. 2). The nuccal hairline was normal and there was no evidence of seborrhea. The eyebrows and lashes were similarly affected. Epicanthal folds were present and there was a chronic conjunctivitis presumably secondary to hair fragment irritation. The eyes were otherwise normal. The remainder of the physical examination was within normal limits. Specifically, the nails were grossly normal and no teeth had yet erupted.

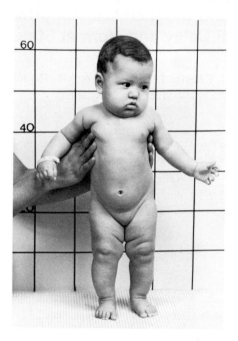

Fig. 1. *Case 1,* age 5 months, demonstrates normal stature and sparse, short hair.

The following laboratory studies were normal: CBC, urinalysis, total serum protein electrophoresis, SGOT, alkaline phosphatase, bilirubin, creatinine, uric acid, BUN, blood glucose, cholesterol, phosphorus, calcium, albumin, total protein and electrolytes. Serum copper was 120 μg% (normal 65-165) and ceruloplasmin was 32 mg% (normal 20-35). Metabolic screening of urine included normal reducing substances, mucopolysaccharides, ferric chloride, carbohydrates, amino acids and there were no consistent atypical peaks by ultraviolet analysis. Serum amino acids were normal.

A complete skeletal survey was negative. Bone age was compatible with the patient's chronologic age. Audiology was compatible with a 4-month level. Im-

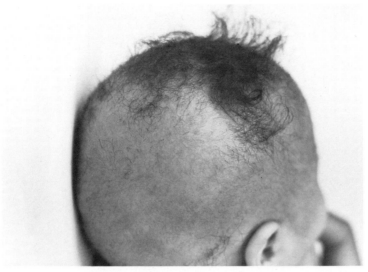

Fig. 2. Brittle hair is absent in areas of pillow contact in *Case 1*. Hair has not been trimmed but is short.

pedance audiometry demonstrated no functional middle ear pathology. A Denver developmental screening test, administered in Spanish by a native Spanish speaking psychologist, suggested a delay of 4 weeks in mental development.

Case 2

This patient, the 5 5/12-year-old brother of the proband, was born 3/25/69 after an uncomplicated gestation. Except for bronchopneumonia at 15 days of age, there were no other neonatal problems. Review of systems and the medical history were otherwise negative. Developmentally, the patient sat at 6 months, walked at 14 months and spoke distinct words at 3 years.

At approximately 1 month of age dry, brittle, fragile hair which fractured easily was first noted. Regrowth appeared to be normal and focal areas of alopecia developed over pressure areas. Hair cuts have never been necessary. Petrolatum hairdressing is used by the parents in an attempt to control the fracturing. The patient plays and relates well with his peers but fatigues easily and generally seems less alert and inquisitive than his 2-year-old brother. There is no history of focal neurologic problems or convulsions. The patient was admitted with his sister for evaluation.

At age 5 5/12 years, physical examination showed a head circumference of 49.5 cm, height 109 cm, U/L ratio 1.14, span 111 cm, and weight 20.4 kg (Fig. 3). The head was normocephalic and facies symmetric. The hair resembled his sister's although it contained a large quantity of lubricant. The dry brittle hair was longest anteriorly and almost absent in the occipital region (Fig. 4). Many broken hairs covered the patient's shirt and pillow. The eyebrows were similarly

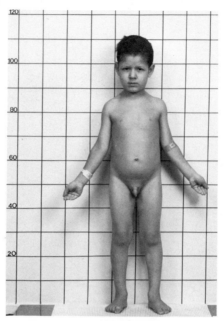

Fig. 3. *Case 2* at 5 5/12 years. Note the sparse eyebrows and scalp hair which has never been cut.

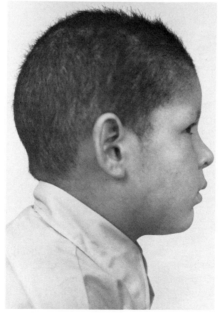

Fig. 4. Dry, brittle, easily fractured hair gives a closely trimmed appearance.

affected. The corneas were clear by slit-lamp examination and ophthalmologic exam revealed mild hyperopic astigmatism and mild congenital tortuosity of the left retinal vessels. The physical examination was otherwise negative, including dermatologic and neurologic evaluation.

The following laboratory values were normal: CBC, urinalysis, total serum protein electrophoresis, SGOT, alkaline phosphatase, bilirubin, creatinine, uric acid, BUN, blood glucose, cholesterol, phosphorus, calcium, albumin, total protein and electrolytes. Serum copper was 139 μg% and ceruloplasmin was 49 mg%. Metabolic screening of the urine revealed slightly positive toluidine blue spot test for mucopolysaccharides. Urine demonstrated normal large weight mucopolysaccharides: breakdown product ratios and a borderline elevation of urinary keratan sulfate (7 mg). Lymphocyte culture revealed a normal 46,XY karyotype by aceto-orcein stain as well as Q-banding. Other urinary metabolic tests including reducing substances, ferric chloride, amino acids, carbohydrates and ultraviolet analysis were normal as were serum amino acids.

The skeletal radiologic survey included a Panorex. Hand and wrist views revealed delayed bone age: Long bones and carpals were consistent with 3 years and phalanges consistent with a 4½-year bone age. Radiographs were otherwise normal. There were no changes of any intrinsic bone dysplasia and no radiographic stigmata of Menkes kinky hair disease. Audiology was normal. Psychometrics, performed in Spanish by an experienced native Spanish speaking psychologist, demonstrated short attention span, poor comprehension with inappropriate answers, short memory and decreased fine motor abilities. The patient's functional mental level was below 4 6/12 years giving an approximate IQ of 75-80.

Hair Analysis

The 2 affected sibs' hair was dry, very brittle and fragile. Hair was collected from the nape of the neck in all family members except in *Case 1* from whom hair was collected from any scalp site available. Light microscopy revealed irregular hair surface and diameter with decreased cuticular layer. Many of the hairs had paintbrush-like fractures at the irregularly spaced nodular swellings which resembled trichorrhexis nodosa (Fig. 5). Indirect illumination revealed irregular birefringence (Fig. 6).

Dr. A. C. Brown of Atlanta, Georgia, kindly provided sulfur analysis (Table 1) and scanning electron microscopy. Scanning electron micrographs, arranged in pedigree style in Figure 7, demonstrated normal morphology in the mother and the clinically normal sib. The father's hair demonstrated a slight decrease in cuticles which is consistent with diminished hair sulfur content (1). The severely defective cuticle formation in the 2 affected sibs was noted. The patient's hair also contained markedly depressed hair sulfur content.

Prior to trace element analysis, hair was prepared by rinsing first with alcohol, then acetone, washing with detergent and finally rinsing with distilled water (2). Trace elements were analyzed by energy dispersive x-ray fluorescence spectroscopy (3) using a secondary target. This equipment was designed by one of the authors (P.S.O.) for the serum trace element analysis (4). Though the results of the trace element analysis in this family (Table 2) are in close agreement with those obtained by conventional atomic absorption spectrophotometry, this new instrumentation needs to be compared further with conventional techniques.

Fig. 5. Irregularly spaced nodular swelling along the hair shaft.

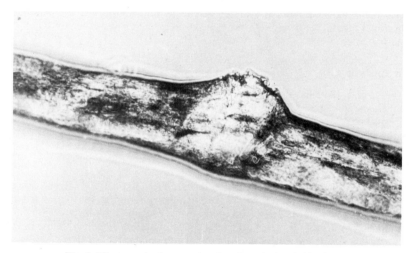

Fig. 6. Microscopic close-up showing the paintbrush-like fracture.

TABLE 1. Hair Sulfur Content (%)

Father	4.40
Mother	4.91
Case 2	2.02
Brother	4.89
Case 1	2.19

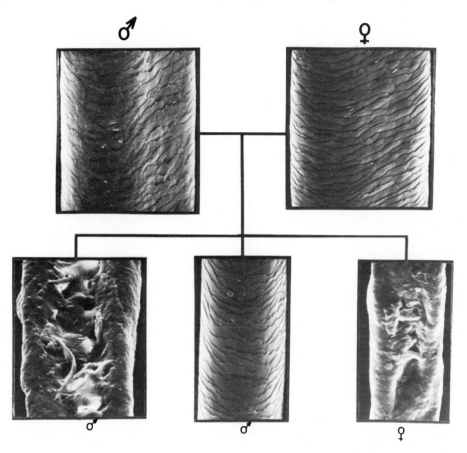

Fig. 7. Scanning electron micrographs of the hair in pedigree arrangement. Note the grossly disturbed cuticular morphology in the 2 affected sibs.

TABLE 2. Hair Trace Element Content (ppm)

	Copper	Zinc	Calcium
Father	16	179	1780
Mother	13	141	803
Case 2	40	356	5670
Brother	16	148	806
Case 1	56	240	3480

DISCUSSION

Atypical hair morphology occurs in association with many syndromes including several with developmental delay or frank retardation (5, 6). Recently, for example, an Amish pedigree has been described with a recessively transmitted syndrome manifest by brittle hair, intellectual impairment and short stature (7, 8). Our patients may be distinct from this syndrome since they both exceeded the 50th percentile in height for their age. In addition comparison of scanning electron micrographs of hairs from the Amish pedigree and our own reveals morphologic differences. The Amish patients reportedly had diminished hair sulfur content with no abnormality of trace metals such as copper and calcium whereas our patients had elevations of trace elements *and* decreased hair sulfur content. We have not, however, analyzed hair proteins in our pedigree to compare with the Amish pedigree. Brown (9) reported a patient who had trichoschisis with diminished hair sulfur and increased hair shaft copper whose hair did not resemble ours morphologically and whose patient lacked developmental delay.

Individual syndromes may demonstrate a variety of dysmorphic hairs (1). Though our patients demonstrated primarily paintbrush-like fractures of trichorrhexis nodosa, their scanning electron micrographs morphologically resembled an 11-year-old retarded female with trichoschisis (10). We are attempting to obtain some of her hair for analysis in our laboratory. Trichorrhexis nodosa usually results from mechanical trauma though it commonly occurs in various syndromes such as argininosuccinic aciduria and Menkes (11). We excluded argininosuccinic aciduria with normal amino acid studies. Menkes disease is ruled out by pedigree data, normal serum copper levels and lack of characteristic pili torti in this syndrome.

Decreased hair sulfur content is a nonspecific finding associated with diminished cystine content and decrease in the cuticular layer in many dysmorphic hair syndromes (10). The affected hair in this syndrome also contained significant elevations of zinc, copper and calcium. This was determined by x-ray fluorescent spectroscopy relative to the unaffected sib and parents whose values were within the standards established for other patients whose hair had been analyzed by different techniques (12, 13).

This x-ray fluorescent system provides several benefits when compared to atomic absorption techniques. Atomic absorption techniques are destructive; they require vaporization of samples for analysis. Our system retains the specimens intact for future reanalysis. Atomic absorption is limited to one element analysis sequentially, whereas the energy dispersive solid state detector employed in this system simultaneously detects x rays emitted by different elements in the sample and distinguishes between x-ray photons of different energies (4). Both techniques allow analysis of samples as small as a few milligrams; however, sample preparation for atomic absorption analysis is more involved and may introduce variability into the results. For concentrations greater than 50 ppm, our system's precision (ie reproducibility) exceeds 98%. The precision with the current pedigree approached 97% for copper and 99% for zinc. These values are comparable or superior to atomic absorption precision (14).

Environmental factors can probably be excluded as the cause for the biochemical variations noted in the patients' hair. The hair defect was noted to be congenital. There is an intermediate-aged sib who has morphologically normal hair and on psychometrics exceeds the average mental functional level for his age group. Finally, the hair sulfur and trace element data in the unaffected members of this pedigree resemble analyses on normal individuals from other series.

The significance of the trace element elevations is as yet unknown (15). One explanation might be a concentration effect though this has not been noted in other morphologic hair defects in which trace elements were analyzed. If the trace elements are selectively bound or localized with certain hair protein, presumably the concentration of trace elements would rise when other proteins (such as the cuticle layer) are absent. The clear association of altered hair trace element values with metabolic disorders such as decreased hair zinc content in zinc deficient dwarfs (16, 17) raises the possibility that pursuit of a metabolic basis or a biochemical handle for this syndrome might be fruitful.

SUMMARY

We have presented 2 affected sibs—a male and female—with unaffected parents and sib from a small remote northern Mexican village. The syndrome includes mental deficit, brittle hair with decreased cuticular layer and an apparently collapsed cortex. The patients' hair contains decreased sulfur content and increased concentrations of trace elements as determined by x-ray fluorescent spectroscopy. Studies are underway to evaluate other apparently similarly affected children from the village where our family originated.

ACKNOWLEDGMENT

We would like to express our appreciation to Dr. Nicola M. DiFerrante for analysis of urinary mucopolysaccharides.

REFERENCES

1. Brown, A. C.: Congenital hair defects. In Bergsma, D. (ed.): Part XII. "Skin, Hair and Nails," Birth Defects: Orig. Art. Ser., vol. VII, no. 8. Baltimore: Williams & Wilkins Co. for The National Foundation-March of Dimes, 1971, pp. 52-68.
2. Ong, P. S. and Cox, H. L.: A line focusing x-ray monochromator for the analysis of trace elements in biologic specimens. Medical Physics.
3. Goulding, F. S. and Jaklevic, J. M.: Photon-excited energy-dispersive x-ray fluorescent analysis for trace elements. Ann. Rev. Nuclear Sci. 23:45, 1973.

4. Ong, P. S., Lund, P. K., Litton, C. E. and Mitchell, B. A.: An energy dispersive system for the analysis of trace elements in human blood serum. Advances in X-Ray Analysis 16:124, 1973.

5. Porter, P. S.: The genetics of human hair growth. In "Skin, Hair and Nails," op. cit., pp. 69-85.

6. Pollitt, R. J. and Stonier, P. D.: Proteins of normal hair and of cystine-deficient hair from mentally retarded siblings. Biochem. J. 122:433, 1971.

7. Jackson, C. E., Weiss, L. and Watson, J. H. L.: Brittle hair with short stature, intellectual impairment and decreased fertility: An autosomal recessive syndrome in an Amish kindred. Pediatrics 54:201, 1974.

8. Watson, J. H. L., Weiss, J. and Jackson, C. E.: Scanning electron microscopy of human hair in a syndrome of trichoschisis with mental retardation. In Brown, A. C. (ed.): "The First Human Hair Symposium." New York: Medcom, 1974, pp. 170–184.

9. Brown, A. C., Belser, R. B., Crounse, R. G. and Wehr, R. F.: A congenital hair defect: Trichoschisis with alternating birefringence and low sulfur content. J. Invest. Dermatol. 54:496, 1970.

10. Brown, A. C., Gerdes, R. J. and Johnson, J.: Scanning electron microscopy and electron probe analysis of congenital hair defects. In: "Scanning Electron Microscopy Proceedings." Chicago: I.I.T. Research Institute, 1971, p. 371.

11. Pollitt, R. J., Jenner, F. A. and Davies, M.: Sibs with mental and physical retardation and trichorrhexis nodosa with abnormal amino acid composition of the hair. Arch. Dis. Child. 43:211, 1968.

12. Eatough, D. J., Christensen, J. J., Izatt, R. M. and Hartley, C.: Level of selected trace elements in human hair. In "The First Human Hair Symposium." op. cit., pp. 377-387.

13. Schroeder, H. A. and Nason, A. P.: Trace metals in human hair. J. Invest. Dermatol. 53: 71, 1969.

14. Cox, H. L.: Personal communication.

15. Goldsmith, L. A. and Baden, H. P.: The analysis of genetically determined hair defects. In "Skin, Hair and Nails," op. cit., pp. 86-90.

16. Hambidge, K. M., Hambidge, C., Jacobs, M. and Baum, J. D.: Low levels of zinc in hair, anorexia, poor growth, and hypogeusia in children. Pediatr. Res. 6:868, 1972.

17. Prasad, A. S., Halsted, H. A. and Nadimi, M.: Syndrome of iron deficiency anemia, hepatosplenomegaly, hypogonadism, dwarfism and geophagia. Am. J. Med. 31:532, 1961.

Familial Aortic Dissection With Iris Anomalies — A New Connective Tissue Disease Syndrome?*

David Bixler, DDS, PhD and Ray M. Antley, MD

Dissecting aneurysm of the aorta is an often fatal disease; its incidence in autopsy studies has been reported to be as high as 1%, although the population prevalence is probably no greater than 1/10,000 (1). The victim is typically a middle-aged male in previously good health except for the presence of hypertension. Occasionally the disease occurs in younger individuals, most often those who have a primary connective tissue disorder, such as Marfan syndrome or Ehlers-Danlos syndrome. Sometimes dissection occurs in association with pregnancy. In general this disease occurs 2 to 3 times more frequently in males than in females. The disease process primarily affects large arteries with a major elastic component in the wall, and is characterized by early accumulations of mucoid material which are lacunar in size and appear between the elastic laminas. These ultimately enlarge to produce loss of both the muscle and the fine elastic fibers. Confluence of these foci leads to a cystic appearance, resulting in the designation given by Erdheim of cystic medial necrosis. The presence of the typical Erdheim lesion has been reported in 50 to 75% of autopsied cases with dissecting aneurysm (1).

The association of dissecting aneurysm with pregnancy appears to be a real one. It has been noted by Schnitker and Bayer (2) that about 50% of dissections in women under age 40 occur in association with pregnancy, most instances being antepartum. Of the 505 cases of aortic dissection (127 females) reviewed by Hirst et al (1), only 15 were associated with pregnancy. However, Schnitker and Bayer (2)

*Publication number 75-18 from the Medical Genetics Department; supported in part by the Indiana University Human Genetics Center (PHS P 01 GM 21054).

Birth Defects: Original Article Series, Volume XII, Number 5, pages 229–234

found that half of their female cases occurred in association with pregnancy (24 out of 49). In 1954 Mandel et al (3) reviewed Schnitker and Bayer's cases, adding 12 of their own, and noted that 36 out of 70 had been associated with pregnancy.

The most common predisposing cause to aortic dissection appears to be hypertension. Hirst et al (1) reported that 30% of the 505 cases reviewed by them had a history of hypertension and 63% had clinical hypertension occurring with the dissection. On this basis one might be tempted to argue that hypertension, as a complicating factor of pregnancy, is the fundamentally related cause in pregnant women with dissections. However, both Schnitker and Bayer (2) and Mandel et al (3) presented evidence that hypertension was no more frequent in the pregnancy-associated cases than in the females over age 40. Thus, pregnancy would seem to initiate a special set of circumstances leading to dissection in a significant proportion of these young women.

It has been suggested that the hormone relaxin, which is present in elevated amounts during late pregnancy and which regulates the general relaxation of ligaments and other joint structures characteristic of this period, may be significantly involved in pregnancy-associated dissections (4). However, dissections have been reported during all 3 trimesters of the antepartum period as well as the postpartum period (2). Furthermore, McKusick (4) has been unable to induce dissection in animals with large doses of relaxin, both with and without pretreatment with vasopressor agents.

Finally, heredity has been invoked as a significant etiologic element in several familial instances of aortic dissection in which Marfan syndrome was discounted but could not be excluded (5–8). A notable exception is the report of Hanley and Jones (7) of 3 persons with dissections in 2 generations of a single family. Their description seems to exclude Marfan syndrome.

The purpose of this report is to describe a family in which 3 women have had aortic dissections, 2 of which occurred 2 weeks' postpartum. None of these persons had stigmata of the Marfan syndrome but each had unique and remarkable eye findings, suggesting that this may be a hitherto undescribed, heritable connective tissue disease complex.

CASE REPORTS

Case 1

The proband in this family was 30 years of age at the time of her death. She was under treatment for about 4 years for intermittent hypertension (systolic pressures of 170-180). She had previously given birth to 3 healthy children, 2 boys and a girl. Eight days after delivery of the fourth child, a girl, she experienced severe back and left arm pain. An aortogram revealed a possible ascending arch dissection and surgical repair was attempted. Postoperatively she had marked central nervous system changes and was areflexic. Death occurred the next day.

Physical findings. The proband was 165 cm tall at the time of death. No U/L ratios were obtained but photographs give no suggestion of skeletal disproportion. No arachnodactyly was present. Both pupils showed irregularities, giving them a colobomatous appearance described as a "fried egg" by the family. This eye condition was present as long as this woman's sister could remember and was noted by several physicians. Joint hyperextensibility was not evaluated. No pectus excavatum or kyphoscoliosis was present.

At autopsy myocardial hypertrophy in an otherwise normal heart with normal coronary arteries was demonstrated. The aortic dissection involved the ascending and descending arches, thoracic and abdominal aorta and both iliac arteries. No mesenteric or renal artery involvement was noted. Histopathologic examination of the aorta revealed degenerative changes of the media with cyst formation typical of that described as Erdheim disease. No coarctation was noted.

Case 2

The mother of the proband had been the picture of health for most of her life and appeared much younger than her actual age. Two years before her death she had a pulmonary embolism with resultant collapse of a lung. She was placed on anticoagulant therapy for the remainder of her life. About the same time, rheumatoid arthritis was diagnosed and treated with cortisone. Specifically, she did not have hypertension at any time during this last period of illness. In 1962 at the age of 56 she had the symptoms and signs of a massive aortic dissection and died within hours. No autopsy was performed. The patient was also known to have an irregularly shaped pupil of one eye which was described as resembling her daughter's eyes. Her height was reported as about 170 cm and although no body segment measurements are available, photographs (Fig. 1, on right) do not suggest skeletal disproportion. No arachnodactyly was present.

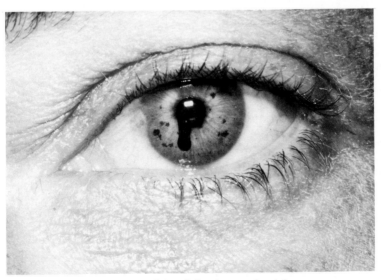

Fig. 1. Left eye of *Case 3* showing pseudocoloboma.

Case 3

Case 3 is the sister of the proband and daughter of *Case 2*. She had had 3 uneventful pregnancies which produced 2 girls and 1 boy, all healthy and without abnormalities. Two weeks prior to delivery of the third child she was admitted for severe chest and back pain. An ECG indicated a myocardial infarction, and SGPT and SGOT enzymes were elevated. She was hospitalized and treated for an infarction. Two weeks later she delivered her third child normally and was discharged after 1 week with a greatly improved ECG. One week later she experienced a sudden onset of back pain, paralysis of legs and vomiting. Her blood pressure was 210/120. She was operated on successfully for a dissecting aneurysm of the abdominal and thoracic aorta and was discharged after 5 weeks.

The dissection was noted to extend from the bifurcation of the iliac arteries upward past the renal arteries into the descending arch (exact extent is still unknown). In numerous areas it was observed that the media had completely separated from the adventitia. No tissue for histopathologic examination was obtained.

At the time of her dissection she was 30 years of age. She is now 43 and in good health.

Physical findings. *Case 3* is 165 cm tall. Her U/L ratio is .98 and she appears normally proportioned. There is no arachnodactyly, joint hyperextensibility or kyphoscoliosis. The skin appears somewhat thin and there is a distinct impression of a lack of subcutaneous tissue. However, no excessive stretchability of skin was demonstrated. The anterior surface of both legs and the lateral ankle areas were, by history, often bumped and easily bruised with prolonged healing time. Several scarred areas were noted but none had the papyraceous or pseudotumor appearance noted in Ehlers-Danlos syndrome. Gorlin sign was absent. The ears were not abnormal in shape or position.

Because she is the only surviving affected patient, detailed comments about her eyes are essential. Clinically, the left eye appears to have a keyhole-like coloboma of the iris (Fig. 1). However, slit-lamp examination reveals a most unusual explanation for this appearance. The pupillary margins show folds of brown pigment, extruding across the margin and out on to the iris. In this eye, the pigment ectropion is most pronounced at 6 o'clock which accounts for the clinical presumption of a coloboma. However, other areas of pigment extruded to a lesser degree were noted. The cause of this ectropion is unknown but we feel it is related to the aortic dissection problem. Further examination suggested atrophy (or hypoplasia) of both the iris stroma and the dilatator muscle. The latter finding was clinically substantiated when the patient stated that after walking into a dark room she could not see for several minutes and had to depend on others to lead her about. Movie theaters have always presented this problem as long as she can remember. Funduscopic examination was within normal limits although more detailed studies of the retina are planned. The sclera was white.

The right eye has no obvious clinical appearance of abnormality. However, under slit-lamp examination pigment cell layer ectropion could be seen. Essentially, both eyes had identical findings although the right eye was of lesser severity. The sclera was white. These variable clinical manifestations in *Case 3* suggest that bilateral eye involvement probably existed in all 3 women.

A number of eye abnormalities, notably ectopia lentis, have been reported in Marfan syndrome. In a few instances a congenital miosis due to iris and dilatator muscle hypoplasia have been observed (9). Furthermore, ectropion of the iris pigment layer has been described in Rieger syndrome (10) and as a familial occurrence (11). It seems clear, however, that the combination of findings noted here are unique and somehow related to aortic dissection.

The family pedigree is shown in Figure 2. The grandmother, who carries this trait, had 3 children: 2 females who were both affected and a male who is currently asymptomatic. For some time now he has been under treatment for hypertension. The deceased proband, *Case 1,* had 4 children, including 2 girls, who were both examined in the Medical Genetics Clinic. No abnormal excretion of hydroxyproline or hyaluronic acid was noted and U/L ratios varied from .99 to 1.36. Funduscopic and slit-lamp examination of their eyes was within normal limits and no unusual joint or skin hyperextensibility was observed. There were no clinically apparent skeletal defects. However, the older boy (age 11) and girl (age 9) were both treated for congenital hip dislocation.

The surviving sister, *Case 3,* has 3 children, all healthy and none of whom have been examined. The brother of *Case 3,* who has hypertension, has 3 boys ranging in age from 11 to 16, all without known medical problems. None of the family members, other than the 3 women described, has unusual appearing eyes.

It is of interest that the brother of *Case 2* died of a heart attack in his late 30s but no autopsy was performed. He had no obvious eye abnormalities. Since it is extremely difficult, if not impossible, to distinguish a myocardial infarction from an aortic dissection on clinical grounds alone, the presence of fatal heart attacks in young persons in this family raises the question of disease diagnosis. Nevertheless, dominant transmission with a sex-limited effect (pregnancy) seems apparent.

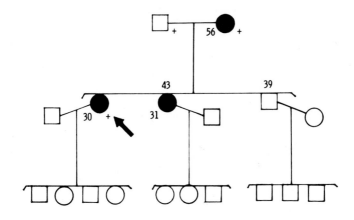

Fig. 2. Family pedigree.

CONCLUSION

It appears that this family has a unique hereditary connective tissue disorder. Plans are underway to evaluate this problem at a biochemical level by studying collagen grown in tissue culture from skin of the proband's sister, *Case 3*. It seems likely that the eye problem described is a part of this dominant disorder but the age of onset remains uncertain. By history it is clinically manifest in childhood, but obviously only a slit-lamp examination can confirm its presence. Detailed studies of the remaining family members are indicated to rule out Marfan or Ehlers-Danlos syndrome, as suggested by other investigators (8, 12).

ACKNOWLEDGMENTS

The authors wish to acknowledge the assistance of Drs. Jackson, McCole and Wilson at the Henry Ford Hospital in Detroit in arranging and conducting examination of *Case 3*.

REFERENCES

1. Hirst, A. E., Johns, V. J. and Kime, S. W.: Dissecting aneurysm of the aorta: A review of 505 cases. Medicine 37:217-279, 1958.
2. Schnitker, M. A. and Bayer, C. A.: Dissecting aneurysm of the aorta in young individuals, particularly in association with pregnancy. Report of a case. Ann. Int. Med. 20:486-511, 1944.
3. Mandel, W., Evans, E. W. and Walford, R. L.: Dissecting aortic aneurysm during pregnancy. N. Engl. J. Med. 251:1059-1061, 1954.
4. McKusick, V. A.: "Heritable Disorders of Connective Tissue," 4th Ed. St. Louis: C. V. Mosby Co., 1974.
5. Von Meyenburg, H.: Ueber spontane Aortenruptur bei zwei Brüdern. Schweiz. Med. Wschr. 20:976-979, 1939.
6. Griffiths, G. J., Hayhurst, A. P. and Whitehead, R.: Dissecting aneurysm of the aorta in mother and child. Brit. Heart J. 13:364-368, 1951.
7. Hanley, W. B. and Jones, N. B.: Familial dissecting aortic aneurysm. Brit. Heart J. 29: 852-858, 1967.
8. Humphries, J. O., Ingle, J. N. and Norum, R. A.: Dissecting aneurysm of the aorta in mother and daughter. In Bergsma, D. (ed.): Part XV. "The Cardiovascular System," Birth Defects: Orig. Art. Ser., vol. VIII, no. 5. Baltimore: Williams & Wilkins Co. for The National Foundation-March of Dimes, 1972, pp. 185-187.
9. Duke-Elder, S. (ed.): Part 2. "System of Ophthalmology," Normal and Abnormal Development: Congenital Deformities. St. Louis: C. V. Mosby Co., 1963, vol. III, p. 1102.
10. Baratta, O.: Alterazione congenite famigliari dell'iride. Boll. d'Oculi 16:339-354, 1937.
11. Falls, H. F.: Gene producing various defects of anterior segment of eye, with pedigree of family. Am. J. Ophth. 32:41-52, 1949.
12. Lichtenstein, J.: Erdheim's cystic medial necrosis in father and son. In "The Cardiovascular System," op. cit., pp. 282-283.

A Possible New Mental Retardation Syndrome

Jessica G. Davis, MD and Charlotte Lafer, MD

Similar physiognostic findings in a sister and brother, suggest that they have a previously undescribed multiple congenital anomaly-mental retardation (MCA-MR) syndrome. The family pedigree (Fig. 1) reveals no consanguinity or other individuals with similar problems. The patient's mother (*III-5*) had a first trimester spontaneous abortion between the birth of the 2 affected children.

CASE REPORTS

Case 1

This patient, a 4½-year-old girl (Fig. 2), was born to a 24-year-old, gravida 0, para 1 white woman, the product of an uncomplicated pregnancy. Labor occurred spontaneously but delivery was by emergency C section because of fetal distress. Apgar score is unknown. Birthweight was 3290 gm. Head circumference was normal.

Physical examination revealed hypotonia, unusual facies, frontal bossing, a beaked nose, large low-set ears with normal contours, bilateral epicanthal folds, partial ptosis of the eyelids, micrognathia, cleft of the soft palate, incurving 2nd and 5th digits bilaterally and hypoplastic toenails. Dermatoglyphics showed a right simian crease, a preponderance of ulnar and radial loops and no arches.

Blood counts, urinanalyses, serum electrolytes, BUN, creatinine, calcium, phosphorus and proteins were all within normal limits. Roentgenograms of the skull and chest showed no abnormalities. ECG was normal. Chromosome analysis revealed no abnormalities.

During her first year of life, this child had 2 bouts of aspiration pneumonia and one urinary tract infection. An IVP and a voiding cystourethrogram showed normal kidneys but vesicourethral reflux. Follow-up urinalyses and urine cultures have been normal.

Chronic problems since birth have been a marked delay in the acquisition of all developmental milestones and failure to thrive. Currently, she is enrolled in a program for retarded children. At age 4½ years, she walks holding on, eats finger foods, says only "Mama" and is not toilet trained. Her height and weight are below the 3rd percentile.

Birth Defects: Original Article Series, Volume XII, Number 5, pages 235–238
© 1976 The National Foundation

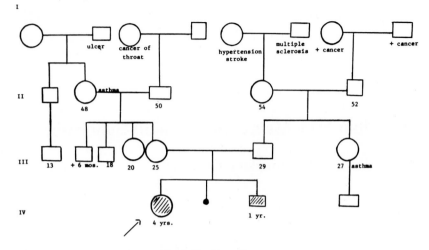

Fig. 1. Family pedigree.

Case 2

The 18-month-old male sib of *Case 1* (Fig. 3) was born following an uneventful pregnancy. He was delivered by elective C section. Birthweight was 2836 gm. Apgar scores were not recorded. Head circumference was normal.

He was hypotonic and was found to have low-set ears, an upturned nose, hypertelorism, bilateral epicanthal folds, micrognathia, cleft of the soft palate, first degree hypospadias with severe meatal stenosis, cryptorchidism, sacral dimple and bilateral dislocated hips. Subsequently, he developed bilateral inguinal hernias.

Blood counts, urinanalyses, urine cultures, serum electrolytes, BUN, creatinine, calcium, phosphorus and proteins were all within normal limits. Roentgenograms of the skull and chest and an IVP were read as normal. Radiographs of his hips showed valgus deformities of both femoral necks. Cytogenetic studies utilizing Giemsa and fluorescent banding techniques revealed apparently normal male karyotypes.

Initially, he exhibited failure to thrive. At present his height and weight are below the 3rd percentile. Development is delayed. At 18 months of age, he sits independently for a few seconds, rolls over, does not crawl and makes only babbling sounds. He is more responsive and relates better to his parents and others than his sister.

This child's medical course has been stormy, requiring 7 hospitalizations for aspiration pneumonia and 9 hospital admissions for bouts of vomiting and dehydration. No evidence of any GI anomaly including tracheoesophageal fistula and pyloric stenosis have been found. EEGs have been read as normal.

SUMMARY

The abnormal clinical features of both patients are summarized in Table 1. The appearance of our patients is reminiscent of patients with the Smith-Lemli-Opitz syndrome (1). However, at birth they were not small and underweight for gestational age. They did not exhibit microcephaly, cataracts, syndactyly, plano-

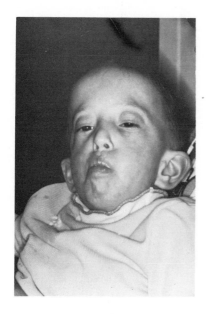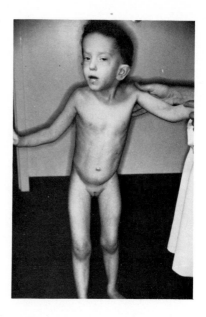

Fig. 2. *Case 1* at 4.5 years.

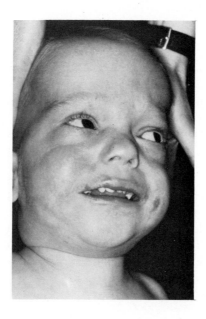

Fig. 3. *Case 2* at 18 months.

Table 1. Clinical Manifestations

	Female Sib	Male Sib
General		
Mental retardation	+	+
Hypotonia	+	+
Normal birthweight	+	+
Subsequent failure to thrive	+	+
Craniofacial		
Head		
Normocephaly	+	+
Frontal bossing	+	+
Nose		
Beaked	+	–
Upturned	–	+
Eyes		
Epicanthal folds	+	+
Cataracts	–	–
Partial ptosis	+	–
Hypertelorism	–	+
Palate		
Cleft soft palate	+	+
Genitalia		
Normal female	+	–
Hypospadias	–	+
Cryptorchidism	–	+
Bilateral inguinal hernias	–	+
Limbs		
Unilateral simian crease	+	–
Hypoplastic toenails	+	–

valgus/metatarus valgus or dorsiflexion of the 1st toes. There was also no history of episodic high-pitched screaming. In addition, our patients did not have the described dermatoglyphic patterns of this syndrome.

Malformations in these sibs were similar to those seen in trisomy 18 (2) or pseudotrisomy 18. However, our patients presented with no evidence of intrauterine growth retardation, congenital heart disease, GU or GI anomalies or flexion deformities of their limbs. Chromosome analyses performed on samples of blood and skin obtained from both children utilizing Giemsa and fluorescent banding techniques revealed apparently normal female and male karyotypes, respectively. Both parents were studied and found to have normal karyotypes.

Therefore, we think that these 2 children present with a MCA-MR syndrome which appears to be due to the homozygous state of an autosomal recessive gene with a 25% recurrence risk.

REFERENCES

1. Smith, D. W., Lemli, L. and Opitz, J. M.: A newly recognized syndrome of multiple anomalies. J. Pediatr. 64:210, 1964.
2. Smith, D. W., Patau, K., Therman, E. and Inhorn, S. L.: The no. 18 trisomy syndrome. J. Pediatr. 60:513, 1962.

Focal Palmoplantar and Marginal Gingival Hyperkeratosis — A Syndrome*

Robert J. Gorlin, DDS

Raphael et al (1) reported the combination of plantar and marginal gingival hyperkeratosis in a kindred involving several generations. In 1974 while visiting Athens, Greece, I had the opportunity to see another family with the same disorder and in 1975 I examined a large kindred in Minneapolis. Other examples were reported by James and Begg (2) and Fred et al (3).

Autosomal dominant inheritance is indicated by transmission of the disorder through several generations. There was male-to-male transmission in all kindreds.

Focal hyperkeratosis of the soles was more marked over the weight-bearing areas: heels, toe pads and metatarsal heads (Fig. 1). Hyperkeratosis of the palms also seemed trauma related (Fig. 2). Hyperhidrosis was noted in the hyperkeratotic areas. The hyperkeratotic areas appeared around puberty in most patients. Involvement of the distal portion of the finger- and toenails with keratin deposits first involved the toes at 4–5 years of age, followed by fingernail changes at 8–9 years of age (Fig. 3).

Sharply marginated hyperkeratosis involved the labial- and lingual-attached gingiva. The hyperkeratotic areas appeared in early childhood and increased in severity with age (Fig. 4).

*This study was made possible by USPHS program grant in Oral Pathology DE-1770.

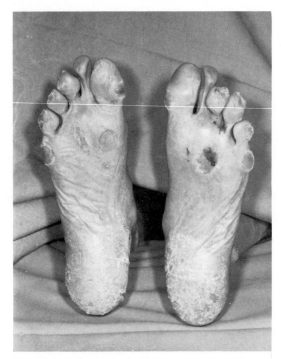

Fig. 1. Focal hyperkeratosis of soles first appearing about puberty.

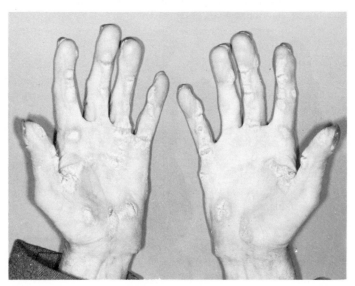

Fig. 2. Focal hyperkeratosis of palms.

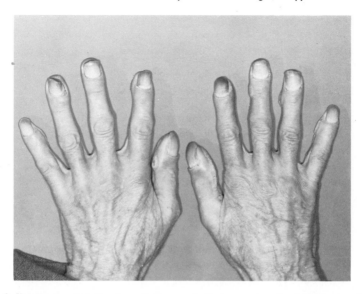

Fig. 3. Keratin accumulation beneath fingernails first appearing about 8–9 years of age.

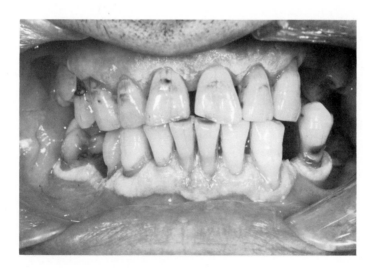

Fig. 4. Sharply marginated hyperkeratosis of fixed gingiva.

REFERENCES

1. Raphael, A. L., Baer, P. N. and Lee, W. B.: Hyperkeratosis of gingival and plantar surfaces. Periodontics 6:118-120, 1968.
2. James, P. and Begg, D.: Tylosis. A case report. Br. J. Oral Surg. 11:143-145, 1973.
3. Fred, H. L. et al: Keratosis palmaris et plantaris. Arch. Intern. Med. 113:866-871, 1974.

Monosuperocentroincisivodontic Dwarfism*

Elizabeth B. Rappaport, MD, Robert Ulstrom, MD and Robert J. Gorlin, DDS

The syndrome consists of growth retardation and a single deciduous and per- manent central incisor. A fine eponym was coined until we attempted to obtain permission from Mr. Burr Tillstrom, the creator of a whimsical TV puppet dra- gon. However, permission to use the eponym was not granted on the grounds that "having a single central incisor was quite normal for a dragon — and besides it was prehensile" — a property not manifested in any of our patients.

Within a one-month period, 6 patients, each exhibiting a single maxillary de- ciduous and permanent central incisor and severe growth retardation, came to our attention (Fig. 1). We have examined 3 of these patients: a 5 5/12-year-old, a 3 1/12-year-old and a 1 1/12-year-old. Information concerning the other 3 pa- tients (one, an adult) came to us in personal communication.

Growth hormone studies were carried out on 3 patients. Insulin, L-dopa and arginine stimulation failed to produce the normal response in the 2 older children. The 1 1/12-year-old responded normally. Personal communication with D. Smith, Seattle (1975) and A. Lucky, Bethesda (1975) confirmed our finding of deficient growth hormone. Other endocrine studies were normal.

Possibly germane are the reports of 2 patients. Francés (1) described an 11-year- old girl with hypopituitary dwarfism and a midline defect of the upper lip, ie the labial or philtral portion of the primary palate was missing. Her nose was depressed

*This study was made possible by USPHS program grant in Oral Pathology DE-1770.

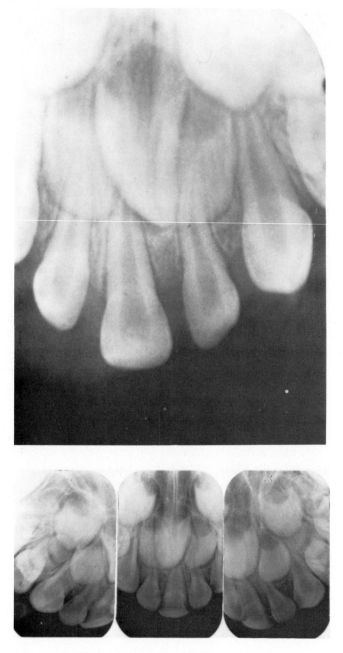

Fig. 1. Single deciduous and permanent central incisors.

due to failure of formation of the nasal bone and cartilage. The sella was hypoplastic. Ocular hypotelorism was apparent in the photographs. Her olfactory sense was intact. Lowry (2) noted that a woman of normal stature with a single maxillary central incisor and ocular hypotelorism gave birth to an infant with holoprosencephaly.

Rather assiduous perusal of the literature has failed to yield an identical case, however, and we are baffled by what seems to be an untoward number of such patients falling within our ken. Perhaps, it is serendipity. However, a single maxillary central incisor is eye-catching, and if present, likely would have been recorded in other cases of growth hormone deficient dwarfs.

The failure of the youngest of our patients to exhibit abnormal growth hormone levels is certainly perplexing and clearly merits close follow-up.

REFERENCES

1. Francés, J. M., Knorr, D., Martinez, R. and Neuhäuser, G.: Hypophysärer Zwergwuchs bei Lippen-Kiefer-Spalte. Helv. Paediatr. Acta 21:315-322, 1966.
2. Lowry, R. B.: Holoprosencephaly. Am. J. Dis. Child. 128:887, 1974.

Studies of Malformation Syndromes in Man XXXXII: A Pleiotropic Dominant Mutation Affecting Skeletal, Sexual and Apocrine-Mammary Development*

Philip D. Pallister, MD, Jürgen Herrmann, MD† and John M. Opitz, MD†

This paper describes a malformation syndrome in a 16-year-old proposita which includes hypoplasia of ulnar rays and of mammary glands, imperforate hymen and apparent absence of apocrine sweat glands. At least 7 other members are known or presumed to be affected in 3 generations of this family.

CASE REPORT

The proposita (*IV-15*, Fig. 1) was born at term to a 20-year-old primigravida after a pregnancy characterized by marked nausea and vomiting during the first few months. Delivery was from a breech presentation and aided by instruments. Birthweight was 2.78 kg and length was 48 cm. The neonatal course and subsequent psychomotor development were normal except for a hymenotomy performed at 1 month of age because of imperforate hymen. Menarche was at 13½ years and menses occur regularly every 28 days with a normal flow lasting 5 days. The patient, an exceptionally intelligent girl, states that she has no body odor, has never required use of deodorants, has a decreased ability to sweat and has no acne. She does produce cerumen.

At 16 years she was evaluated at the Boulder River School and Hospital because of a congenital abnormality of the left forearm (Fig. 2). Height was 174 cm (95th%), weight 50 kg (70th%) and head circumference 53.4 cm (50th%). There was an apparent mild left craniofacial hypoplasia. The skull was slightly brachycephalic and the midface relatively wide. Inner and outer canthal distance was 31 and 81 mm, respectively, and there was slight mongoloid slanting

*Paper no. 1865 from the University of Wisconsin Genetics Laboratory

†Supported by USPHS/NIH grant GM 20130

Birth Defects: Original Article Series, Volume XII, Number 5, pages 247–254
© 1976 The National Foundation

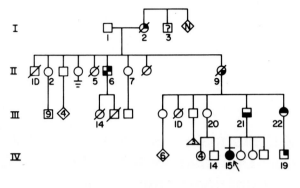

Fig. 1. Pedigree of the proposita's family.

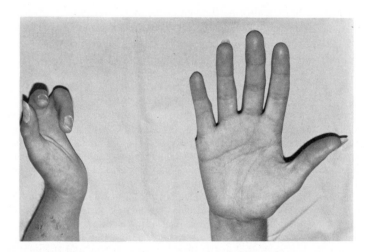

Fig. 2. Absence of ulnar ray structures on left and hypoplasia on right.

of the palpebral fissures. Eyes and nose were normal. The mesodistal diameter of the maxillary first incisors was perhaps slightly increased (9 mm) and all 4 canines were congenitally absent. The arch of the mandible was narrow and the mandibular teeth were crowded. The bony palate was high but not narrow, and the uvula was bifid. The auricles were well differentiated; however, the left helix was tightly furled in its upper portion. The lower portions of the helices were hypoplastic bilaterally.

Developmental asymmetry was more pronounced with regard to the chest and the arms. The neck and upper back showed about 10 junctional nevi on the left and 5 on the right. The left clavicle was somewhat thinner than the right one and the xiphisternum was absent. The areolas and nipples were small and glandular breast tissue was flat and 4 cm in diameter on the right and 5 cm on the left (patient has had mammary implants). Abdomen and back were unremarkable. Axillary and pubic hair were sparse; the configuration of the pelvis was gynecoid. Pelvic examination showed a tight hymenal ring and, under general anesthesia for hymenotomy, the uterus was judged to be relatively small and the cervix to be immature.

Both upper limbs showed hypoplasia of forearm bones and abnormal development of ulnar rays, considerably more marked on the left than on the right. The left 4th and 5th fingers and metacarpals were absent. The 2nd and 3rd fingers were of normal size but there was (partially repaired) 50% cutaneous syndactyly between them. There was marked camptodactyly of the thumb which was subluxated at the MP joint. The longitudinal palmar flexion crease was hypoplastic and only one (distal) transverse palmar flexion crease was present on the left. The dermatoglyphics included a radially displaced *a* triradius and absence of the *b*, *c*, *d* and *t* triradii. A radial loop was present on the thumb and the 3rd finger, and an ulnar loop on the index finger. The fingernails were hyperconvex with leukodysplasia. The left forearm was severely shortened (only 12 cm long) and there was severe limitation of pronation and supination although complete extension was possible. The left antecubital flexion crease was absent. Cubitus valgus of about 30° was present bilaterally.

On the right hand, the patient showed evidence of surgically corrected ulnar hexadactyly with a rudimentary stump of a 6th digit arising at the base of the 5th finger. The 5th metacarpal was short, the 5th finger was tapered and hyperextensible in the PIP joint and the hypothenar cushion was hypoplastic. There was mild camptodactyly in the MP joints of the 2nd through 5th fingers. The 1st metacarpal was short. There were well-formed dermatoglyphics with an increased number of white lines and parallelity of the transverse palmar creases. There was a *t″* and properly placed *a*, *b*, *c* and *d* triradii and a radial carpal loop in the hypothenar area. An ulnar loop was present on all the right fingertips except for a tented arch on the 5th finger.

The length of both humeri (left 33.5 cm, right 32 cm) was within normal limits, but both ulnas were short [left 12 cm, right 25 cm (10th%)].

The lower limbs were normal. A distal loop was present in each hallucal area. Neurologic examination was unremarkable except for a decreased left brachioradialis reflex.

Roentgenograms confirmed the clinical findings (Fig. 3). The left ulna was severely hypoplastic and the left radius was correspondingly deformed and showed marked bowing. The left humerus was slightly hypoplastic. The olecranon was developed as a separate ossification center and there was articulation between the distal humerus and the proximal end of the hypoplastic ulna rather than with the olecranon. The radial head was dislocated. The hand skeleton lacked the 4th and 5th fingers and metacarpals, and the triquetrum, hamate and pisiform bones. Articulations of the structures present were apparently normal. The distal phalanges of the 1st, 2nd and 3rd fingers were small. The right humerus was slightly hypoplastic. In addition this patient had a moderately severe, S-shaped thoracic scoliosis; IVP demonstrated slight posterior rotation of the left kidney.

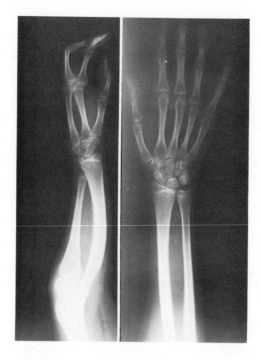

Fig. 3. Roentgenogram of forearms at 16 years.

Family History

The pedigree of the family is outlined in Figure 1. The following historic information is available on other family members who were not examined:

I-2: had "stiff and crooked" 5th fingers.

I-3: "affected."

II-6: This man had "deformed and almost jointless" 5th fingers; photos demonstrated delayed physical and sexual maturation with striking sexual (facial profile and beard) maturation during his 20s.

II-9: Patient's paternal grandmother is reported as not visibly affected but is said to have had no body odor and decreased sweating.

III-21: Patient's father had right 5th finger camptodactyly which was amputated in PIP joint in childhood; left hand reportedly normal; said to have had late sexual maturation with delayed growth spurt (height now about 188 cm) and not to produce body odor.

III-22: Patient's aunt had congenital absence of left 5th finger; hymenotomy was performed at 16 years; breast is absent on right and hypoplastic on left but nipples are present bilaterally; no breast enlargement during pregnancy or by hormonal therapy; had C section for either cephalopelvic disproportion or uterine insufficiency; has had mammary implants.

IV-19: Patient's cousin has small 5th fingers fixed in extension.

DISCUSSION

The principal clinical findings of the proposita are summarized in Table 1. The condition is of genetic and pathogenetic interest. Apparently it is a previously undescribed dominantly transmitted condition.

Goldenring and Crelin (1) reported a 25-year-old woman and her daughter with bilateral congenital absence of breasts and nipples. The mother also had a depressed nasal bridge, highly arched palate, sparse axillary and pubic hair, and in biopsies from the axilla, numerous active eccrine but no apocrine glands. The teeth, bony thorax, pectoralis muscles and all other findings were reported as normal and it is presumed the limbs were also normal. Her 3½-month-old daughter also had marked hypertelorism, a highly arched palate and depressed nasal bridge; an older daughter was unaffected. The mother had no observable breast development during her 2 pregnancies. Fraser (2) reported another family with dominant amastia which included absence of the nipples and normal limbs.

A sporadic case with skeletal and urogenital abnormalities similar to those of our patient was described by Muechler (3). This was a 26-year-old woman with slight right ulnar hypoplasia and marked left ulnar rays hypoplasia including absence of fingers 4 and 5, and syndactyly of fingers 2 and 3 (with a single

TABLE 1. Summary of Findings in the Proposita and Some of Her Relatives*

Upper Limbs:
    Left — Absence of 4th and 5th fingers, metacarpals, triquetrum, hamate and pisiform bones with secondary abnormalities of palmar flexion creases and dermatoglyphics; shortness and malformation of forearm bones (ulna > radius) with limitation of pronation and supination and cubitus valgus; 50% cutaneous syndactyly fingers 2 and 3; camptodactyly of thumb with subluxation at MP joint and shortness of 1st metacarpal; slight hypoplasia of distal phalanges fingers 1−3.

    Right — Postaxial poly(hexa)dactyly with shortness of 5th metacarpal, tapering and hyperextensibility at PIP joint of 5th finger and hypoplastic thenar cushion; mild camptodactyly at MP joint of fingers 2−5; shortness of 1st metacarpal.

    (Other family members: Stiff, small, crooked or absent 5th fingers.)

Mammary-Apocrine Glands:
    Absence of body odor and acne (also in paternal grandmother).
    Hypoplasia (or absence) of mammary gland, hypoplasia of areolas and nipples (also in aunt).

Vertebral Column: Moderately severe thoracic scoliosis.

Teeth, Palate: Increased diameter of maxillary first incisors; congenital absence of canine teeth; high palate, bifid uvula.

GU System: Slight malrotation left kidney; imperforate hymen (also in aunt).
    (Delayed sexual maturation in father of proposita and his maternal uncle.)

*Data within parentheses relates to relatives of the proposita.

distal phalanx extending off the joined proximal and middle phalanges). This patient also had an abnormal spinous process of the first sacral segment, abnormal bony spurs of the ileum, absence of the left kidney and a low-lying malrotated right kidney. The vagina ended blindy at a depth of 1–2 cm. She lacked the uterus and the oviducts except for their infundibular segments. The ovaries appeared normal. There was normal development of the breasts and external genitalia. We are unable to determine at this time whether this patient has a somewhat atypical and rather severe form of the condition which is present in our patient and members of her family. In our patient's family the breasts and nipples were present but hypoplastic and the limbs were abnormal. Male-to-male transmission was not observed; inheritance, therefore, could be either autosomal or X chromosomal. Expressivity was quite variable.

This dominantly inherited multiple congenital anomaly (MCA) syndrome shows apparent mosaic pleiotropy affecting several tissues, organs and body regions, ie upper limbs, the mammary-apocrine glands, vertebral column, teeth, palate and GU system. For an association such as ulnar ray hypoplasia and ipsilateral absence or hypoplasia of the mammary glands and axillary apocrine glands, we suggest the term *polytopic anomaly*. This term refers to etiologically related noncontiguous anomalies which may all be related to a single underlying pathogenetic mechanism as a single developmental field complex (DFC) or single developmental field anomaly (DFA). The presumption that this may be a DFC would be confirmed by occurrence of the same complex in an etiologically different syndrome, and/or a teratologic or developmental genetic demonstration of how the several components of the DFC relate to each other pathogenetically.

Analogous examples might be the concurrence of radial and cardiac malformations in the Holt-Oram syndrome; chest, costal, mammary and hand malformations in the Poland anomaly; radial ray hypoplasia and anemia in the autosomal recessive Fanconi anemia and TAR syndrome, and in the autosomal dominant TW syndrome (4) (ulnar-radial rays hypoplasia with aplastic anemia and/or leukemia); and forelimb and thumb anomalies in certain patients with hydrocephalus or myelodysplasia. The concurrence of acral and renal malformations in a large number of etiologically distinct syndromes and the experiments of Lash (5), which show how such anomalies may be developmentally related and dependent on each other, make it also possible to refer to some of these acrorenal associations as probable polytopic DFCs. We think it is important to make a distinction between polytopic DFCs (such as the acrorenal DFCs) and monotopic DFCs (such as the alobar holoprosencephaly DFC) since presumably different types of (relational) pleiotropy are involved.

It is postulated, on the basis of regional occurrence, that some of the manifestations of our patient may represent a polytopic anomaly, ie those involving the breasts, limbs and ipsilateral axillary apocrine glands. It is presently not understood how they are related pathogenetically. However, with respect to the possible limb-breast-apocrine gland polytopic anomaly, it might be postulated that because of a primary effect on the embryonal mammary ridge, which in about 10 mm human embryos extends from the origin of the upper limb bud to the ipsilateral lower limb bud, mammary and axillary apocrine gland hypoplasia and ipsilateral hypoplasia of the upper limb may be the result of a single pathogenetic event or chain of events. The embryologic development of axillary apocrine odor glands has not been clearly traced; however, their normal distribution (and their absence in our patient) corresponds strikingly to the course of the embryonal mammary ridge (6). We therefore suggest that our patient has a hypoplastic mammary ridge defect as part of a polytopic mammary ridge-limb anomaly which, in turn, is a component anomaly of a dominantly inherited mosaic pleiotropic trait that may additionally affect vertebral, tooth, palate and urogenital sinus development and, at time in males, gonadal function. In this connection it would be of great interest to determine if patients with the Poland anomaly, especially those with ipsilateral breast aplasia or hypoplasia, have an apocrine gland deficiency in that axilla.

This condition is of additional interest since it is another MCA syndrome in which duplications or excesses of tissue (ie anomalies of abnormal development) may be manifestations, in the same patient or different members of the family, of hypoplasias or deficiencies of the corresponding tissue (ie anomalies of incomplete development). A triphalangeal thumb (rarely preaxial polydactyly) is a well-known alternate manifestation of thumb-radius hypoplasia or aplasia; in the thalidomide syndrome it was one of the mildest expressions of what, with an increased dose and longer exposure to the drug, was frequently represented by a severe phocomelia-like defect. The hands of our patient demonstrate a similar but much rarer phenomenon involving the ulnar side, with one hand showing postaxial hexadactyly, the other absence of the 4th and 5th rays and corresponding carpal bones. In this syndrome the lower limbs are not affected, but we have seen a similar occurrence in upper and lower limbs, ie polydactyly of hands with oligodactyly and/or deficiency of leg bones; or in different segments of the same limb, ie deficiency of leg bones with polydactyly of feet. The TAR syndrome seems to be an exception in that severe mesomelic and at times rhizomelic deficiency of the upper limbs is apparently never associated with oligodactyly of hands. The short rib-polydactyly syndromes and the Meckel syndrome illustrate the opposite phenomenon — unusual polydactyly of hands apparently never associated with intercalary deficiencies.

ACKNOWLEDGMENTS

We appreciate the cooperation of Drs. Warren M. Swager (Sheridan, MT), J. Kent Boughn (Helena, MT) and James E. Nickel (Helena, MT) in providing clinical information.

REFERENCES

1. Goldenring, H. and Crelin, E. S.: Mother and daughter with bilateral congenital amastia. Yale J. Biol. Med. 33:466–467, 1961.
2. Fraser, F. C.: Dominant inheritance of absent nipples and breast. In "Novant anni delle leggi mendeliane." Roma: Instituto Gregorio Mendel, 1956, pp. 360–362.
3. Muechler, E. K.: Müllerian duct agenesis associated with renal and skeletal abnormalities. Am. J. Obstet. Gynecol. 121:567–568, 1975.
4. Herrmann, J., Gilbert, E. and Opitz, J. M.: Dysplasia, malformations and cancer, especially with respect to the Wiedemann-Beckwith syndrome. Presented at an NICHD/NIA and NCI Workshop on Regulation of Cell Proliferation and Differentiation in May 1975 at the Institute for Medical Research, Camden, N. J. Proc. being edited by W. W. Nichols and R. W. Miller.
5. Lash, J. W.: Normal embryology and teratogenesis. Am. J. Obstet. Gynecol. 90:1193–1207, 1964.
6. Arey, L. B.: "Developmental Anatomy," 7th Ed. Philadelphia: W. B. Saunders, 1965.

Hallux Syndactyly — Ulnar Polydactyly — Abnormal Ear Lobes: A New Syndrome

Michael J. Goldberg, MD and Hermine M. Pashayan, MD

Polydactyly and syndactyly are relatively common congenital malformations present both as isolated abnormalities and as part of syndromes. Syndactyly shows definite predilection for the 1st postaxial web space (1). The axis of the foot traverses the 2nd toe; that of the hand passes through the 3rd digit. In most cases of syndactyly of the foot, the 2nd and 3rd toes are united; in the hand, the 3rd and 4th digits are webbed. Isolated cutaneous syndactyly of the great toe (hallux) and 2nd toe is rare. Syndromes which include syndactyly of the 1st and 2nd toe are the acrocephalosyndactylies (2), acropectorovertebral dysplasia (3), craniofacial dysostosis (4), popliteal pterygium syndrome (5) and orofacial-digital syndrome (5); but these conditions invariably have, in addition, multiple other congenital anomalies, both of the feet (eg tarsal and metatarsal fusions, complex bony syndactyly of all toes) as well as other distinguishing features such as mental retardation, popliteal webbing, complex bony syndactyly of the hand, vertebral and sternal anomalies and craniosynostosis. Sporadic cases of cutaneous syndactyly of the hallux and 2nd toes have been reported associated with finger syndactyly (6), polydactyly of the feet (7) and congenital hallux varus with short 1st metatarsal (8).

The distinguishing feature of a complete cutaneous syndactyly of the 1st and 2nd toes drew attention to this previously unreported syndrome of malformations. There seems to be 3 definite aspects to this condition involving very specific anatomic areas (ie the lobe of the ear, the ulnar border of the hand and the preaxial border of the foot). The spectra of anomalies are:

Ears: The presence of either a deep horizontal groove (Fig. 1) or a nodule on the lobe (Fig. 2).

Birth Defects: Original Article Series, Volume XII, Number 5, pages 255–266
© 1976 The National Foundation

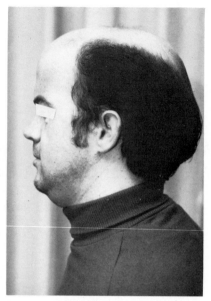

Fig. 1. *II–2*: Typical deep horizontal grooves in the earlobe.

Fig. 2. *II–3*: Characteristic nodule on the earlobe.

Hands: Ulnar polydactyly with the 6th digit invariably originating from the ulnar aspect of the 5th finger just distal to the metacarpophalangeal joint and ranging from a soft tissue nubbin (Fig. 3) to a more complete finger, including phalanges and fingernail. No metacarpal or other skeletal abnormality in the hand is present.

Feet: Four areas of malformation are present. 1) Partial-to-complete cutaneous syndactyly between the great and 2nd toes (Fig. 4) and occasionally between the 2nd and 3rd toes in addition (Fig. 5). 2) Preaxial polydactyly, usually consisting of a broad (Fig 6) or bifid (Fig. 7) distal phalanx of the great toe or 2 separate phalanges (Fig. 8), but occasionally a completely separate preaxial toe of distal phalanx and nail (by history). 3) A separate delta phalanx located at the medial aspect of the 1st metatarsophalangeal (MTP) joint (Fig. 9), often fused with the medial condyle of the proximal phalanx (Fig. 10). 4) A hypoplastic but definite metatarsal located proximally between the normal 1st and 2nd metatarsal (Figs. 11 and 12).

INDEX CASE

A 3-month-old white male (*III-7*), the product of an uneventful full-term pregnancy, was noted to have ulnar polydactyly and toe syndactyly at birth. The bilateral pedunculated digits, composed of phalanx and fingernail, were surgically excised in the nursery. Examination revealed a healed scar at the proximal ulnar aspect of the proximal phalanx of the small fingers. The ears exhibited nodules on the lobe (Fig. 13). There was mobile, complete, cutaneous syndactyly between the great and 2nd toes which were of the same length. Except for a small umbilical hernia, the remainder of his physical examination was normal. Roentgenographs of the hands were normal. Roentgenograms of the feet revealed absence of the ossification center of the distal and middle phalanges of the 5th toe (Fig. 14).

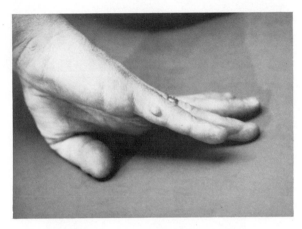

Fig. 3. *II—3*: Ulnar polydactyly. Soft tissue nubbin on ulnar border of 5th finger. No previous surgery.

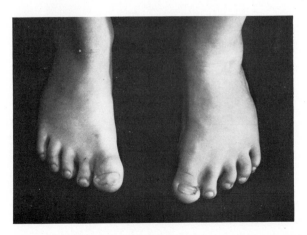

Fig. 4. *III—5*: Three-year-old female with near-complete cutaneous syndactyly between 1st and 2nd toes bilaterally. Broad great toenails.

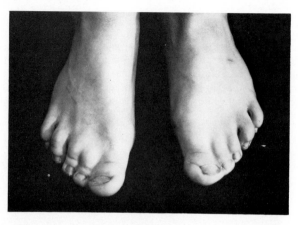

Fig. 5. *III–4*: Five-year-old male exhibiting complete cutaneous syndactyly between 1st, 2nd and 3rd toes bilaterally.

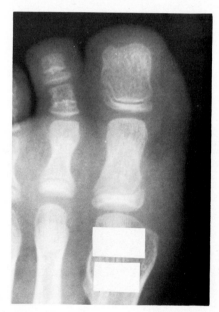

Fig. 6. *III–2:* Seven-year-old female with markedly broad distal phalanx of the great toe on the left.

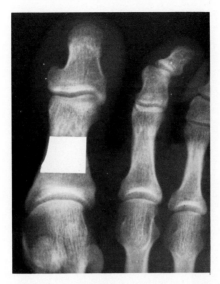

Fig. 7. *II—2*: Bifid distal phalanx of the great toe present bilaterally.

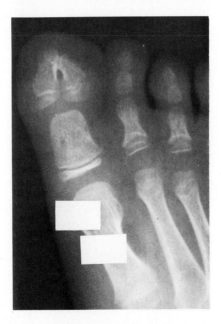

Fig. 8. *III—5*: Three-year-old female with complete duplication of the distal phalanx of the great toe on the right only.

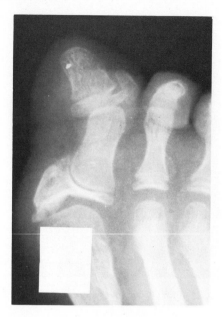

Fig. 9. *III—2*: Seven-year-old female with delta phalanx with L-shaped epiphyseal plate at the MTP joint of the great toe. Also, duplication of the distal phalanx.

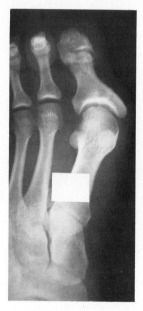

Fig. 10. *II—3*: Delta phalanx fused to proximal phalanx creating an extended abnormal medial condyle. Also, broad distal phalanx and accessory hypoplastic metatarsal.

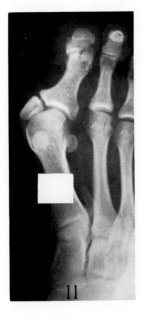 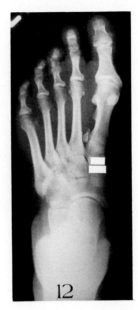

Fig. 11. *II—3*: Hypoplastic accessory metatarsal located proximally between normal 1st and 2nd metatarsals. Also, separate delta phalanx at MTP joint and duplication of distal phalanx.

Fig. 12. *II—1*: Accessory hypoplastic metatarsal. Also, fused deltal phalanx creating prominent medial condyle of the proximal phalanx. Note small exostosis on lateral aspect of 1st metatarsal.

SUMMARY OF AFFECTED FAMILY MEMBERS

The abnormal physical findings of the 10 affected family members are summarized in Figure 15 and Table 1. The 56-year-old paternal grandmother (*I—1*) had a small horizontal groove in the left earlobe. Her hands were normal. There was cutaneous syndactyly between the great and 2nd toes as far as the base of the nail; a broad distal phalanx of the great toe on the left and a duplicated distal phalanx on the right; a prominent medial condyle of the proximal phalanx on the right; and a rudimentary metatarsal between the 1st and 2nd metatarsals on the right.

The 33-year-old paternal aunt (*II—1*) had a short horizontal groove in the lobes of both ears. She had ulnar polydactyly of both hands, treated surgically with skin tags remaining. Her feet demonstrated minimal webbing between the 2nd and 3rd toes on the left foot. There had been bilateral preaxial polydactyly with surgical removal during childhood of an extra toe, described as having a distal phalanx and a toenail. An accessory delta phalanx and an extra metatarsal between the 1st and 2nd metatarsal were also present bilaterally. A small exostosis on the lateral aspect of the 1st metatarsal was present only in this patient (Fig. 12).

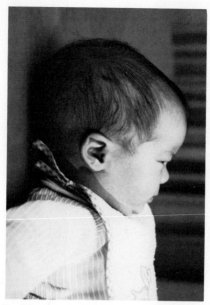

Fig. 13. *III–7*: Index case with nodule on the earlobe.

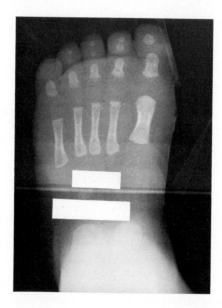

Fig. 14. *III–7*: Index case. Note absences of ossification centers of phalanges of 5th toe.

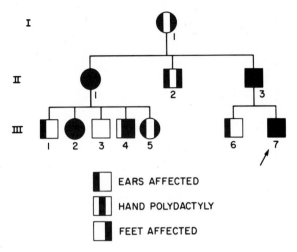

Fig. 15. Pedigree of the syndrome showing an autosomal dominant pattern of inheritance.

The 30-year-old paternal uncle (*II−2*) had a horizontal groove on the lobe of the left ear (Fig. 1). His hands were normal. There was no syndactyly, but a bifid distal phalanx of the great toe (Fig. 7) and a hypoplastic accessory metatarsal were present bilaterally.

The 28-year-old father of the proband (*II−3*) had nodules on his earlobes (Fig. 2). His hands had soft tissue nubbins on the ulnar aspect of the 5th finger (Fig. 3). He had complete cutaneous syndactyly between the 1st and 2nd toes bilaterally; duplication of the distal phalanx of the great toe bilaterally; delta phalanx at the MTP joint bilaterally (separate on the right, fused on the left); and the accessory hypoplastic metatarsal was present bilaterally (Figs. 10 and 11).

The 9-year-old first cousin (*III−1*) had a deep groove on the left earlobe, but otherwise had normal hands and feet except for a very minimal web between the 2nd and 3rd toes on the left.

The 7-year-old first cousin (*III−2*) demonstrated horizontal grooves on the earlobes bilaterally, and a history of ulnar polydactyly bilaterally consisting of digit and fingernail which was surgically excised at age one year. Her feet had cutaneous syndactyly between the 1st and 2nd toes bilaterally and partial syndactyly between the 2nd and 3rd toes. There was history of surgical excision of a preaxial toe bilaterally at one year of age. A broad distal phalanx was present on the left (Fig. 6) and duplication of the distal phalanx of the great toe on the right. The delta phalanx at the MTP joint was present on the right but not the accessory metatarsal (Fig. 9).

The 6-year-old first cousin (*III−3*) had normal ears, hands and feet, although the web space between the 2nd and 3rd toes on the left was not as deep as on the right foot.

A 5-year-old first cousin (*III−4*) had normal ears. His hands were described as having ulnar polydactyly with the extra finger including a fingernail being removed in the nursery, although distinct skin tags remained. His feet demonstrated complete cutaneous syndactyly between the 1st, 2nd, and 3rd toes bilaterally

TABLE 1.

Pedigree number	Age	Sex	Ear		Hand ulnar polydactyly	Hallux syndactyly	Feet			Additional findings
			Groove	Nodule			Preaxial polydactyly	MTP delta phalanx	Accessory metatarsal	
I-1	56	F	– L	– –	– –	R L	R L	R –	R –	minimal web toe 2–3 L
II-1	33	F	R L	– –	R L	– –	R L	R L	R L	
II-2	30	M	– L	– –	– –	– –	R L	– –	R L	
II-3	28	M	– –	R L	R L	R L	R L	R L	R L	
III-1	9	M	– L	– –	– –	– –	– –	– –	– –	partial syndactyly toe 2–3 L
III-2	7	F	R L	– –	R L	R L	R L	R –	– –	partial syndactyly toe 2–3 L R
III-3	6	M	– –	– –	– –	– –	– –	– –	– –	? partial web toe 2–3 R
III-4	5	M	– –	– –	R L	R L	R L	– –	– –	complete syndactyly toe 2–3 L R
III-5	3	F	– L	– –	– –	R L	R –	– –	– –	
III-6	2	M	R L	– –	– –	– –	– –	– –	– –	
III-7	3/12	M	– –	R L	R L	R L	– –	– –	– –	absent middle phalanx toe 5 L R

(Fig. 5). Broad distal phalanges of the great toe bilaterally was the only x-ray abnormality.

A 3-year-old first cousin (*III−5*) had a deep groove in the left earlobe. Her hands were normal. There was cutaneous syndactyly between the 1st and 2nd toes bilaterally (Fig. 4). Complete duplication of the distal phalanx of the great toe on the right side was the only abnormal x-ray finding (Fig. 8).

The proband's 2-year-old brother (*III−6*) had ears with a deep groove on the lobe bilaterally. His hands and feet were normal.

DISCUSSION

The kindred spanned 3 generations and of the 11 family members studied, 10 demonstrated some or all of the abnormalities. The triad of ear, hand and foot malformations was seen in 4 patients; ear and foot malformations were seen in 3 patients; isolated earlobe abnormalities were seen twice; and one patient had only the hand and foot affected. Abnormal earlobes were present in 9 of the 10 affected patients. The feet were affected in 8. Ulnar polydactyly was present in 5 cases.

The hallux syndactyly and preaxial polydactyly of the foot were apparent at birth. However, the extra hypoplastic metatarsal was present only in the adults studied and the accessory delta phalanx was present in only one older child (*III−2*). It is suggested that these latter malformations are cartilaginous during childhood and ossify with maturity.

To our knowledge the paternal grandmother was the first to show manifestations of the syndrome. All 3 of her children showed some (*II−2*), or all (*II−1*) and (*II−3*) of the features of the syndrome. In turn, 6 out of 7 of her grandchildren showed some or all of the manifestations. This pedigree is compatible with autosomal dominant inheritance since male-to-male transmission has occurred in 2 instances.

SUMMARY

A new syndrome is described. Its distinguishing features are hallux syndactyly, ulnar polydactyly and abnormal earlobes with other roentgenographic abnormalities along the medial border of the foot. It is inherited as an autosomal dominant.

REFERENCES

1. Kelikian, H.: "Congenital Deformities of the Hand and Forearm." Philadelphia: W. B. Saunders, 1974, p. 331.
2. Temtamy, S. and McKusick, V. A.: Synopsis of hand malformations with particular emphasis on genetic factors. In Bergsma, D. (ed.): Part III. "Limb Malformations," Birth Defects: Orig. Art. Ser., vol. 5, no. 3. White Plains: The National Foundation - March of Dimes, 1969, pp. 125−184.

3. Grosse, F. R., Herrmann, J. and Opitz, J. M.: The *F*-form of acro-pectoro-vertebral dysplasia: The *F*-syndrome. In "Limb Malformations," op. cit., pp. 48–63.
4. Dodge, H. W., Jr., Wood, M. W. and Kennedy, R. L. J.: Craniofacial dysotosis: Crouzon's disease. Pediatrics 23:98, 1959.
5. Pfeiffer, R. A.: Associated deformities of the head and hands. In "Limb Malformations," op. cit., pp. 18–34.
6. Kelikian, H.,: "Congenital Deformities of the Hand and Forearm," op. cit., p. 345.
7. Ibid, p. 423.
8. Zimbler, S.: Personal communication.

X-Linked Syndrome of Congenital Ichthyosis, Hypogonadism, Mental Retardation and Anosmia

Jane C. S. Perrin, MD, Judi Y. Idemoto, MSW, Juan F. Sotos, MD, William F. Maurer, MD and Arthur G. Steinberg, PhD

Congenital ichthyosis is heterogeneous, but rarely found in regular combination with secondary hypogonadism, mental retardation, and impaired ability to smell. In reviewing the coincidence of ichthyosis and other abnormalities, several cases were found to be of particular interest. Sjögren and Larsson (1) cite Rud's reports of 1927 and 1929 of a man and woman with ichthyosis and hypogonadism who were of normal intelligence. DeSanctis and Cacchione in 1932 reported 3 brothers from a first-cousin mating who had pigmented xeroderma with testicular hypoplasia and mental retardation. In 1957 Sjögren and Larsson described 10 males and 18 females with an autosomal recessive condition of congenital ichthyosis, spasticity, mental retardation, and sometimes atypical retinitis pigmentosa (1).

In 1960 Lynch et al (2) reported 4 generations of a kindred in which 3 males, related through females, were shown by examination to have ichthyosis and hypogonadism. Two deceased males related to the former 3 through females were reported to have been affected. The pattern of inheritance was probably sex-linked recessive. Affected males were eunuchoid, with small genitalia, atrophy and azospermia on testicular biopsy, and had ichthyosis verified by skin biopsy. Urinary FSH was low while 17-KS were normal; buccal smears were sex-chromatin negative. Intelligence was not recorded.

Sotos (3, 4) studied 6 males, aged 1–31 years, in 2 generations of a family with ichthyosis, micropenis, small testes, mental retardation and anosmia. Urinary gonadotropins were absent and 17 hydroxysteroid levels low; release of HGH and THs was normal. We have restudied this kindred extensively for genetic counseling and hormone therapy and have extended the pedigree.

Birth Defects: Original Article Series, Volume XII, Number 5, pages 267–274
© 1976 The National Foundation

MATERIALS AND METHODS

This 6-generation Mexican-American kindred (Fig. 1) includes some 250 living members, the majority of whom live in the area comprised of northwestern Ohio, northeastern Indiana and southeastern Michigan. Of a total of 114 males, 20 are affected; 18 are alive; and 16 of these, ranging in age from newborn to 36 years, have been examined by us (*V-80* and *V-81* live in Texas and were unavailable).

Xg^a typing and G-6-PD typing was carried out on blood samples of several individuals, including affected males, and selected obligate female carriers, spouses, and daughters and normal sons of obligate carriers. These same persons underwent testing (age permitting) for color vision with 12 color cards, and olfactory testing with 5 nonirritating solutions: wintergreen, lemon, peppermint, cinnamon and clove. Chromosome karyotype, including Giemsa-staining, was done on 3 affected males and 2 female carriers.

The following endocrine studies were performed on 6 affected males, ages 1, 5½, 11½, 16, 30 and 31 years: plasma immunoreactive growth hormone levels under insulin and arginine stimulation, T_4 and RAI before and after TSH; SU-4885 tests and plasma cortisol response to insulin, urinary gonadotropins, plasma immunoreactive FSH and LH before and after clomiphene, urinary steroids before and after ACTH; and clinical and plasma testosterone responses to the administration of human chorionic gonadotropin (HCG) in those of postpubertal age.

Mental retardation was diagnosed by school report of performance on psychologic testing, or by history and observation.

RESULTS

Physical characteristics of affected males comprised a constant pattern (Table 1), none of which was observed in normal brothers. At birth the skin was bright red but developed fine scales on the trunk and limbs within a few days of life. From infancy through adulthood, scales persisted and became larger and darker with age (Fig. 2A). Involvement was most prominent on the trunk, scalp and distal limbs (none on palms or soles) and was worse in winter. On skin biopsy *(V-15, V-23)* features were characteristic of X-linked ichthyosis with marked hyperkeratosis, acanthosis of the stratum mucosum, prominence of rete ridges and perivascular infiltrate (Fig. 2B). Partial remission of ichthyosis was achieved in those patients using topical vitamin A acid.

Prepubertal males (Fig. 3A) had a normally formed but small penis, a small scrotum and usually undescended abdominal or inguinal testes even after attempted orchiopexy. Untreated affected males of postpubertal age were of normal height and retained a youthful appearance with eunuchoid habitus; absent axillary, pubic and facial hair; luxuriant scalp hair; and high voice. The

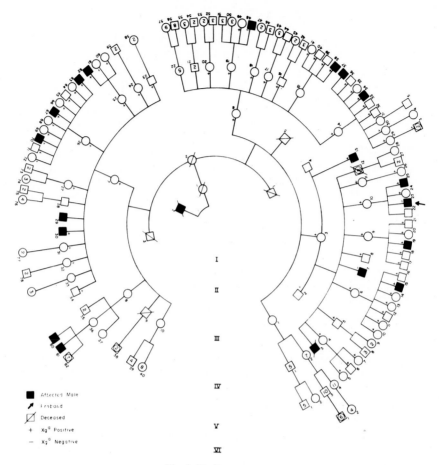

Fig. 1. Pedigree.

penis was small (2 to 5 cm long) and few erections were experienced. Testes were small, 2 cm if palpable. There was no expressed interest in sexual activity and none were married. One 7-year-old genetic male had undergone phallectomy and orchiectomy for sex conversion in infancy.

All had an early history of developmental lag and often hyperactive behavior. The adults either did not attend school or had multiple failures. Younger boys were in special education classes. By testing or observation, these males were functioning in the educable mentally retarded to borderline range (IQ 50–80). There was no history of school failures or special education in normal male sibs.

TABLE 1. Clinical Characteristics

	No. affected/No. examined
Ichthyosis	16/16
Body hair (before testosterone)	0/16
Small penis	15/16
Testes undescended, small	15/16
Anosmia	14/14
Color blind	0/14
Mental retardation or borderline	15/15
Other	
hypoplastic kidney	2
spasticity	1
hypotonia	1
deafness	1
Low plasma FSH, LH	6/6
Decreased ACTH reserve	6/6
Xg^a type	
positive	11
negative	3

Of the 15 affected males old enough to test for smell and vision, none were color blind, but all failed to detect and identify 3 or more of the 5 odors. Normal brothers had normal olfactory response. Chromosome karyotype was 46,XY in 3 affected males tested. None of the female obligate carriers had ichthyosis or impaired olfactory sense as a clinical marker. All conversed in a manner demonstrating normal intelligence except 2 who were emotionally disturbed, supported by psychiatric history. Female offspring of obligate carriers who were of school age had no history of school failures.

Hypogonadotropic hypogonadism was demonstrated in the 6 males who underwent endocrine studies. Urinary gonadotropins and plasma FSH and LH before and after clomiphene were low. Testicular biopsy in the 30-year-old showed absence of Leydig cells and immature tubules (Fig. 4A). Adequate androgen effects and testicular enlargement resulted from 12 weeks of HCG treatment to the 16-year-old (Fig. 3B) and the biopsy showed Leydig cells and arrested spermatogenesis (Fig. 4B).

Decreased ACTH reserve was judged by SU-4885 testing, plasma cortisol response to insulin-induced hypoglycemia, and urinary steroids before and after ACTH. Secretion of pituitary hormones HGH, TSH and ADH was normal.

All tested individuals were G6PD type B positive; all tested individuals were Xg^a positive except 3 of the affected males.

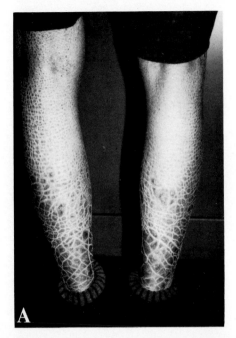

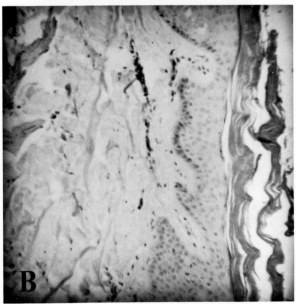

Fig. 2. Skin. A) Ichthyosis. B) Biopsy.

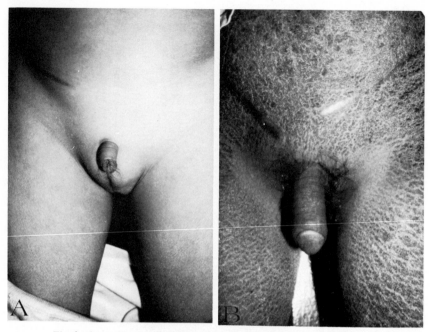

Fig. 3. Genitalia. A) Untreated male. B) Male treated with HCG.

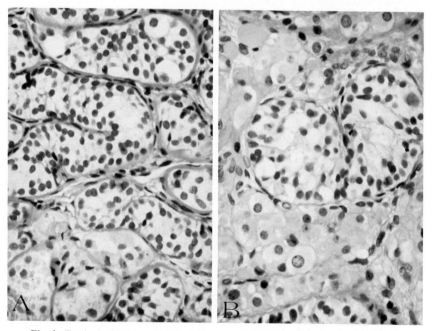

Fig. 4. Testicular biopsy. A) Before treatment. B) After treatment with HCG.

DISCUSSION

X-linked recessive ichthyosis as a single entity is a recognized dermatologic disease (5), but a kindred with an X-linked syndrome of ichthyosis, gonadotropin deficiency, mental retardation, and anosmia is rare. Gene linkage on the X chromosome has been demonstrated between the ichthyosis (isolated entity) and Xg loci (6).

Individuals in the present pedigree with a plus or minus sign above their symbols were tested for Xg^a. The possibility of linkage between this locus and the locus for the ichthyosis syndrome was investigated with the aid of a computer program (LIPED) (7). Xg^a gene frequencies for the British population were used: 0.659 Xg^a positive and 0.341 Xg^a negative; gene frequency for the rare ichthyosis syndrome was taken as 0.0001, normal 0.9999. The likelihood (lod score) of a genotypic configuration at 2 loci for various values of the recombination fraction (theta) was calculated. A low value for theta indicates infrequent recombination and possible linkage between 2 loci; a value of .50 indicates random recombination (no linkage). If a lod score has a negative value, linkage is unlikely. Lod scores are listed in Table 2.

Close linkage is clearly excluded and the trend of the lod scores indicate that the 2 loci (ichthyosis syndrome and Xg^a) probably sort independently.

TABLE 2.

Theta	Lod score
.01	−7.53714
.05	−3.60524
.10	−2.11047
.20	−0.88657
.30	−0.35848
.40	−0.09582

ACKNOWLEDGMENTS

We wish to thank Dr. Henri Frischer for the G6PD typing and Dr. Jon Lochner for dermatologic evaluation.

REFERENCES

1. Sjögren, T. and Larsson, T.: Oligophrenia in combination with congenital ichthyosis and spastic disorders. Acta. Psychiatr. Scand. (Suppl. 113):32 5–112, 1957.
2. Lynch, H. T., Ozer, F., McNutt, C. W. et al: Secondary male hypogonadism and congenital ichthyosis: Association of two rare genetic diseases. Am. J. Hum. Genet. 12:440–447, 1960.

3. Maurer, W. F. and Sotos, J. F.: Sex-linked familial hypogonadism and ichthyosis. *Program* and *Abstracts,* Soc. Ped. Res., 39th Ann. Mtg. 181, 1969.
4. Sotos, J. F.: The endocrine system. In Goodman, R. M. (ed.): "Genetic Disorders of Man," Boston: Little Brown, 1970, pp. 787–789.
5. Wells, R. S. and Kerr, C. B.: Genetic classification of ichthyosis. Arch. Dermatol. 92:1–6, 1965.
6. Kerr, C. B., Wells, R. S. and Sanger, R.: X-linked ichthyosis and Xg groups. Lancet 2:1369–1370, 1964.
7. Ott, J.: Estimation of the recombination fraction in human pedigrees: Efficient computation of the likelihood for human linkage studies. Am. J. Hum. Genet. 26: 588–597, 1974.

Congential Microcephaly, Hiatus Hernia and Nephrotic Syndrome: An Autosomal Recessive Syndrome*

Lawrence R. Shapiro, MD, Peter A. Duncan, MD, Peter B. Farnsworth, MD and Martin Lefkowitz, MD

In 1968, Galloway and Mowat (1) reported 2 sibs with congenital microcephaly, hiatus hernia and nephrotic syndrome. They suggested a possible genetic basis for these conditions, but no other similar sibships have been described until now.

A second family has been studied in which the first 2 children (a female and a male) were found to have congenital microcephaly, hiatus hernia and nephrotic syndrome. There is an unaffected male child. The father is of English ancestry and the mother Italian. There is no consanguinity and the parents are healthy. The pregnancies with both affected children were uneventful with no infections and no exposure to drugs or radiation.

CASE REPORTS

Case 1

The first born child was female. The birthweight was 2240 gm and she was 45 cm long. The head circumference was 28.5 cm as compared to the chest circumference of 29.5 cm. The skull appeared small but of normal configuration and the facies was not unusual. The ears were large and floppy and there was slight redundancy of the posterior cervical skin folds (Fig. 1). The corneas of both eyes were opacified and the lateral aspects of the iris were hypoplastic. Pupillary membrane filaments were noted and these findings were felt to represent failure of cleavage of the anterior chamber of the eye (2).

The significant laboratory data were limited to albuminuria (1—4+) and microscopic hematuria. A chest x-ray film revealed a well-defined pocket of air above the right diaphragm crossing the midline to the left and representing a hiatus hernia. These findings were confirmed by contrast x-ray studies.

*This paper was supported in part by grants from The National Foundation—March of Dimes and The New York State Department of Health Birth Defects Institute.

Birth Defects: Original Article Series, Volume XII, Number 5, pages 275—278

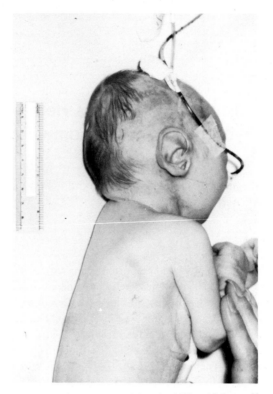

Fig. 1. Large and floppy ears of female child at 13 days of age.

Vomiting occurred on the second day of life and became projectile and unresponsive to conservative measures. At one week of age surgery was performed, and a hiatus hernia was found with most of the stomach and part of a loop of colon in the chest. The hernia was repaired but reexploration for intestinal obstruction was necessary 3 days later. The child died of respiratory complications at 14 days of age.

Postmortem examination revealed that the eyes had prominent vacuolar changes and degeneration of lens fibers posteriorly. The ciliary bodies and irides were underdeveloped. The kidneys revealed microcystic dysplasia (3) and focal glomerulosclerotic changes accounting for the nephrotic syndrome (Fig. 2).

Case 2

The second born child was male. The birthweight was 2708 gm. A small head circumference of 30.5 cm was noted at birth and vomiting began with the first oral feedings. At 5 days of age, albuminuria (2+) was noted and chest x-ray film revealed a hiatus hernia later confirmed by barium studies (Fig. 3). He was managed conservatively and by 5½ months of age, he weighed 3777 gm and was

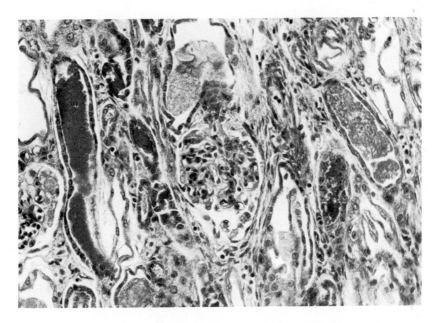

Fig. 2. A glomerulus with prominent mesangial cells incompletely filling Bowman's space. There are thin basement membranes and plump epithelial cells and segmental sclerosis. A hyperplastic juxtaglomerular zone is interposed between the glomerulus and an ectatic distal tubule. A luminated protein cast is present in the lumen. Other tubules contain red cell and granular casts. (H and E, × 245)

59 cm long with a 33 cm head circumference. He was obviously microcephalic and had large, floppy, low and posteriorly set ears similar to his affected female sib (Fig. 1). The eyes were normal with no corneal opacification. A significant degree of micrognathia was noted.

Developmental milestones were delayed and head control was poor.

Laboratory data were not significant except for albuminuria (3–4+). Chromosome analysis, including the Giemsa-banding technique, was entirely normal. No aminoacidemia or aminoaciduria was detected. At 16 months of age, the head circumference measured 34 cm showing very little growth and marked microcephaly. The child was profoundly retarded and had seizure activity controlled with phenobarbital. At 16 months of age, the BUN and creatinine levels remained normal and the urinary specific gravity was normal. Marked albuminuria continued with a drop in total serum protein (albumin 1.9 gm% and globulin 3.0 gm%). Serum cholesterol rose to 341 mgm%.

At 2 years of age, a renal needle biopsy showed the same microcystic dysplasia as described in his deceased sib. By 2 to 3 years old, gross edema was present; he died at 3 years of age due to respiratory infection. Postmortem examination was not permitted.

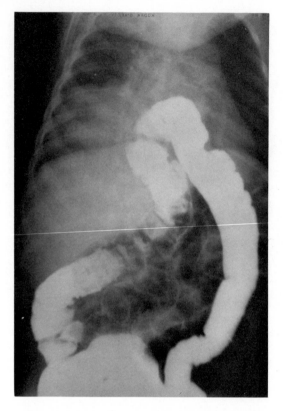

Fig. 3. Contrast study demonstrating hiatus hernia of male child.

DISCUSSION

Both affected sibs were obviously microcephalic at birth and the microcephaly became more marked with age. Both children had large hiatus hernias which were symptomatic at birth. Albuminuria was observed in the first days of life, and microcystic dysplasia and focal glomerulosclerosis accounted for the nephrotic syndrome.

The two separate sibships described above demonstrate that when one of the components of this syndrome occurs congenitally, the possible coexistence of the other features should be investigated, and, if found, the autosomal recessive mode of inheritance appreciated.

REFERENCES

1. Galloway, W. H. and Mowat, A. P.: Congenital microcephaly with hiatus hernia and nephrotic syndrome in two sibs. J. Med. Genet. 5:319, 1968.
2. Reese, A. B. and Ellsworth, R. M.: The anterior chamber syndrome. Arch. Ophthalmol. 75:307, 1966.
3. Oliver, J.: Microcystic renal disease and its relation to "infantile nephrosis." Am. J. Dis. Child. 100:312, 1960.

Polydactyly With Triphalangeal Thumbs, Brachydactyly, Camptodactyly, Congenital Dislocation of the Patellas, Short Stature and Borderline Intelligence

Burhan Say, MD, Eugene Feild, MD, James G. Coldwell, MD, Larry Warnberg, MA and Metin Atasu, PhD

The purpose of this communication is to present a family in which the mother and 3 of her daughters had triphalangeal thumbs associated with other skeletal abnormalities as well as borderline intelligence.

CASE REPORTS

The proband will be described in detail; the findings in other affected family members are shown in Table 1.

Case 1

The proband, a 17-year-old female, presented with recurrent dislocations of the right patella causing problems with her knees since birth. At birth the patient was found to have 6 fingers bilaterally without thumbs. She also had 8 toes on the right foot and 7 on the left. No great toes could be identified while the medial toes originated from the same metatarsal bone. The patient was subjected to multiple surgeries to remove extra toes on both feet along with the metatarsals. Bilateral pollicization in the hands was required for proper function. Her problems with the knees started soon after she began to walk and at age 6 she had bilateral capsular knee surgery to correct patellar dislocations without apparent success. Two years ago the left patella was removed because of recurrent difficulties. This year she had the Hauser operation to preserve the right patella by shifting it into a more normal axis of rotation and length. Family history revealed that 2 of her sisters and the mother were similarly affected (Fig. 1). The patient had 2 older brothers and a sister who were not affected. Although there was a history of a stillborn male, no information was available with regard to the existence of any malformations in this patient. The mother was 23 years old and the father was 27 at the time of her birth. Consanguinity between

Birth Defects: Original Article Series, Volume XII, Number 5, pages 279—286
© 1976 The National Foundation

the parents was denied. The mother stated that there was no history of drug intake or exposure to radiation during her pregnancies which apparently had been quite uneventful. The patient weighed 3500 gm at birth and according to the mother there was nothing unusual about her development except that she could not walk alone until she was 2 years of age.

TABLE 1. Polydactyly With Triphalangeal Thumbs, Brachydactyly, Camptodactyly, Congenital Dislocation of the Patellas, Short Stature and Borderline Intelligence

Clinical Findings	Case 1	Case 2	Case 3	Case 4
Age	17	18	13	40
Sex	F	F	F	F
Height (sm)	146	148	134	150
Polydactyly (preaxial) of the fingers and toes	+	+	+	+
Triphalangeal thumbs	+	+	+	+
Brachydactyly	+	+	+	+
Camptodactyly	+	+	+	+
Congenital dislocation of the patellas	+	+	+	+
IQ	83	85	77	ND
Others:				
Pes planus	+	+	+	+
patent foramen ovale	−	−	+	−
fused labia minora	−	−	+	−
absent lumbar lordosis	ND	ND	ND	+

ND − Not determined

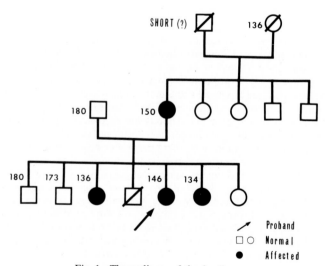

Fig. 1. The pedigree of the family.

The physical examination showed that she was short in stature (146 cm) and had unusual facial features; she had a round face with a prominent nose and a wide nasal bridge (Fig. 2). The mouth had a somewhat carp-like appearance. The neck was short. She also had short limbs associated with brachydactyly (Fig. 3). Old scars were noted bilaterally between the triphalangeal thumb and index finger, along the medial borders of the feet and on the knees. All the fingers were short and showed distal tapering with mild camptodactyly specifically involving the 2nd, 3rd and 4th fingers and clinodactyly involving the 5th finger bilaterally. She had faint interphalangeal lines and no obvious hypothenar and thenar

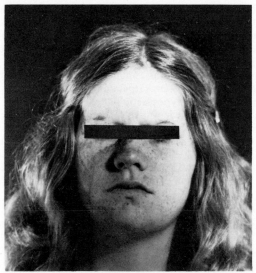

Fig. 2. *Case 1.*

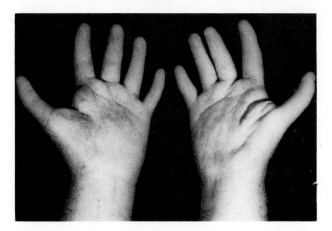

Fig. 3. The hands of *Case 1.*

eminences. The fingers which were left to function as thumbs were triphalan-geal and showed ulnar deviation. Their metacarpal bones were slightly adducted but with satisfactory rotation. In general, she seemed to be able to use her hands for most normal activities. Finally, she had broadening of the wrists bi-laterally. Examination of the knees revealed dislocation of the small right patella which moved laterally in full extension. Some crepitus beneath the patella was noted but no evidence of chondromalacia could be detected.

Laboratory findings including routine urinalysis, blood counts, serum Ca, P, alkaline phosphatase determinations as well as chromosome analysis were normal. Dermatoglyphic studies briefly showed that in the hands the axial triradii were absent while there was a T^u triradius bilaterally. The A basic lines were directed to the radial side and a whorl pattern was seen over the 5th inter-digital space. There were ulnar loops on 7 of the fingers while the remainder 3 had whorl patterns.

X-ray studies showed that there was a splinter-like ossification between the proximal ends of the 1st and 2nd metacarpal bones bilaterally. The middle extra phalanx in the thumb was hypoplastic in both hands (Fig. 4). In the left knee the patella was absent, while it was small and misshapen in the right knee.

Psychometric testing using Wechsler Adult Intelligence Scale showed the pa-tient had an overall functioning in the dull-normal range. Her verbal IQ was 79; her performance IQ was 90 with a full-scale IQ of 83. Pattern profile analysis kindly performed by Dr. Poznanski and his associates revealed a pattern which was similar to that seen in Holt-Oram syndrome (Fig. 5).

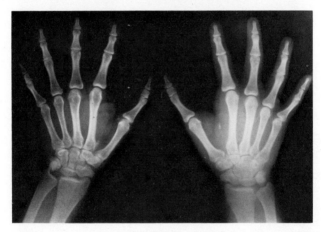

Fig. 4. Radiographs of the hands of *Case 1*.

Case 2

This patient was an 18-year-old female who had almost identical deformities as her younger sister. She had many painful problems with recurrent dislocations mainly of the left patella. Although she complained of no pain, marked chondro-malacia was noted in the right knee. She was also subjected to repeated surgeries for the removal of 2 extra preaxial toes bilaterally as well as bilateral polliciza-tion.

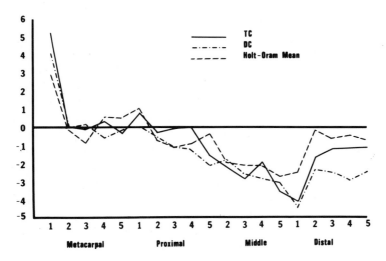

Fig. 5. Pattern profile analysis in *Case 1* and *Case 2*.

Case 3

This 14-year-old female had the least severe knee deformity. However, a complete aplasia of the right anterior tibialis was present and a hypotrophic left anterior tibialis was noted. In addition to the findings shown in Table 1, she was found to have ptosis of the right eyelid, a limited extension of the left arm over the elbow and a cubitus valgus deformity which was more obvious on the left. She was also found to have a 2/4 systolic murmur best heard over the 3rd and 4th intercostal space at left without radiation. Cardiac catheterization at the age of 5 showed the existence of patent foramen ovale without shunting. Finally, she had a fusion of the lower two thirds of the labia minora at birth which was separated easily.

Case 4

The 40-year-old mother of the children described above showed identical findings in her hands, feet and knees. She could not come to the hospital for a detailed examination since she had only limited use of her legs in spite of multiple knee surgeries. Radiographs of the spine showed absence of the normal lumbar lordosis.

DISCUSSION

The family showed a combination of abnormalities which, to the best of our knowledge, have not been previously reported. In addition to quadriaxial polydactyly in all 4 affected members, thumbs and great toes were replaced by triphalangeal digits. They all had similar associated malformations such as characteristic facial features, congenital dislocation of the patellas which were hypoplastic and high-riding, short stature and borderline intelligence (Table 1).

The occurrence of 3 phalanges in the human thumb is a rare deformity. The results of 2 different surveys indicate that its prevalence is about 1 in 25,000 (1, 2). It is frequently familial and, as shown in Table 2, on many occasions triphalangeal thumbs constitute one of the components of various syndromes, including Holt-Oram syndrome, thalidomide embryopathy, congenital hypoplastic anemia and others. Among the conditions listed in the table, there is only one entity as reported by Easton and McKusick in which affected members displayed certain features somewhat resembling those seen in our patients (3). They described 4 individuals in 3 generations of a single family which were found to have preaxial polydactyly of the feet and syndactyly. The cardinal clinical features observed in 3 of the 4 affected family members included 5 triphalangeal fingers without obvious thumbs and malformations of the lower leg in which the tibia was hypoplastic and the fibula was thickened and was displaced on the femur. Interestingly, the patella was also absent. A recent publication on a similarly affected family indicates that the condition is inherited as a dominant trait (4). It should be noted, however, that the short stature in the 3 patients reported was the result of lower limb deformity and they all had normal intelligence. Their facial appearances were not reported as unusual.

Congenital dislocation of the patellas is a well-established entity and is often familial and bilateral (5). A brief review of the literature indicates that it is usually not accompanied by other anomalies.

It is interesting to note that congenital dislocation of the patella is more often seen in females. In Goldwaith's series, it occurred almost entirely in girls or women (6). In a more recent publication the female:male ratio was found to be 3:1 (7).

TABLE 2. Conditions Associated With Triphalangeal Thumbs

Familial triphalangeal thumb (with or without polydactyly)
Cardiomelic (Holt-Oram) syndrome
Thalidomide embryopathy
Congenital hypoplastic anemia
Hypoplasia or aplasia of the tibia, fibular dysplasia and
  polydactyly
Radial dysplasia/Imperforate anus/Vertebral anomalies syndrome
Others
  Juberg-Hayward syndrome
  Mirror hand syndrome
  Ectodermal defects, deafness and mental retardation
  Lobster-claw hand and polydactyly
  Imperforate anus and deafness
  Brachycamptodactyly, congenital dislocation of the patellas, short
    stature and borderline intelligence

A definite familial tendency seems to be present in recurrent dislocation of the patella. Bowker and Thompson (7) found 12 of their 48 patients had similarly affected relatives with multiple generation involvement in 10 of these families. Carter and Sweetnam (8) have described such a family in which 3 patients in 3 generations were found to have recurrent dislocation of the patella associated with joint laxity.

A distinct malformation syndrome was reported by Norvig (9) in which polydactyly and syndactyly of the feet was associated with brachysyndactyly of the hands, acrocephaly and, interestingly, also with displaced patellas. Warkany et al (10) reported on a 16-year-old boy who had syndactyly of the hands with polysyndactyly of the feet in association with acrocephaly and congenital dislocation of the patella. None of these patients had the other findings seen in the affected family members of this report while our patients did not have any evidence of early closure of the cranial sutures. All of these indicate that our patients represent a different syndrome. On the other hand, the association of various forms of polydactyly with congenital dislocation of the patellas (with or without significant lower limb malformations) suggest a common teratologic mechanism. The fact that finger pattern analysis showed similar patterns in 2 of our patients as seen in cardiomelic syndrome further support this hypothesis. Studies of thalidomide embryopathy indicate that the effect of the mutant gene in affected members of this family should have taken place at about the 50th day of intrauterine life, since triphalangeal thumbs have been observed in infants after maternal thalidomide ingestion around this time (11).

The mode of inheritance in this new entity appears to be dominant. Involvement of 3 of the daughters but none of the sons would be consistent with an X-linked dominant as well as with an autosomal dominant inheritance. The follow-up of this family may shed further light on this matter.

REFERENCES

1. Lapidus, P. W., Guidotti, F. P. and Coletti, C. J.: Triphalangeal thumb. Report of six cases. Surg. Gynecol. Obstet. 77:178–186, 1943.
2. Poznanski, A. K., Garu, S. M. and Holt, J. F.: The thumb in the congenital malformation syndromes. Radiology 100:115–129, 1971.
3. Eaton, O. and McKusick, V. A.: A seemingly unique polydactyly syndrome in four persons in three generations. In Bergsma, D. (ed.): Part III. "Limb Malformations," Birth Defects: Orig. Art. Ser., vol. V. no. 3. White Plains: The National Foundation – March of Dimes, 1969, pp. 221–225.
4. Yujnovsky, O., Ayala, D., Vincitorio et al: A syndrome of polydactyly-syndactyly and triphalangeal thumbs in three generations. Clin. Genet. 6:51–59, 1974.
5. Mumford, E. B.: Congenital dislocation of the patella. Case report with history of four generations. J. Bone Joint Surg. 29:1083–1086, 1947.
6. Goldwaith, J. E.: Slipping or recurrent dislocation of the patellae: With the report of eleven cases. Boston Med. Surg. J. 150:169, 1904.

7. Bowker, J. H. and Thompson, E. B.: Surgical treatment of recurrent dislocation of patella. J. Bone Joint Surg. 46A:1451–1461, 1964.
8. Carter, C. and Sweetnam, R.: Familial joint laxity and recurrent dislocation of the patella. J. Bone Joint Surg. 40B:664–667, 1958.
9. Norvig, J.: To tilfaelde of acrocephalosyndatyli hos soskende. Hospitalstiende 72: 165, 1929.
10. Warkany, J., Frauenberger, G. S. and Mitchell, A. G.: Heredofamilial deviations (I-the Laurence-Moon-Biedl syndrome). Am. J. Dis. Child 53:455, 1937.
11. Lenz, M. W.: Chemicals and malformations in man. In "Congenital Malformation." Papers and Discussions Presented at the Second International Congress. Session VI. New York: The International Medical Congress, Ltd., 1964, pp. 263–276.

Recurrent Dislocation of the Patella Versus Generalized Joint Laxity

Steven D. Shapiro, MS, Ronald J. Jorgenson, DDS
and Carlos F. Salinas, DDS

Recurrent dislocation of the patella has been reported to be a manifestation of generalized joint laxity and to be an independent autosomal dominant trait (1). This report will deal with several instances of recurrent patella dislocations in a family with generalized joint laxity in 4 generations.

CASE REPORT

The proposita was a white female, the first of 2 children of a 24-year-old woman and her 26-year-old husband. The pregnancy was full-term and uncomplicated. Birthweight was 3005 gm. The only abnormality noted at birth was a turning in of the feet.

She was seen at the Medical University of South Carolina at 20 years of age with a chief complaint of recurrent dislocation of the patellas. She had had difficulty with loose jointedness of the knees since early childhood. At 9 years of age her left tibial tubercle was transplanted to correct recurrent dislocation of the left patella. A postoperative statement included the opinion that she may have had the Ehlers-Danlos syndrome, although other manifestations of the syndrome were not documented. At 16 years of age the right tibial tubercle was transplanted to correct recurrent dislocation of the right patella. The repaired left knee began to dislocate again during the next year and a transplant of the vastus medialis was performed. This surgery was not successful and 6 months later a patellectomy of the left knee was performed. Subsequent to the patellectomy, surgery was performed to repair a luxating quadriceps aponeurosis. Despite the surgery, both knees still dislocated.

Extensive scars on the inner surfaces of each leg were noted on the physical evaluation. A greater than normal extension was judged to be present at the knees, elbows and fingers. There were no parchment-like scars over the elbows and knees, no elasticity of the skin, no excessive bruising and no abnormalities of the cardiovascular, skeletal or ocular systems.

Birth Defects: Original Article Series, Volume XII, Number 5, pages 287–291
© 1976 The National Foundation

Family History

The patient was of German and Irish extraction. She reported that her parents were not consanguineous and that there was no similarity of last names in their families. Her mother was alive, in good health and had no relatives who had experienced recurrent joint dislocation. Her father was alive and in good health. He was reported to have had recurrent dislocations of his patellas. One of his brothers was reported to have had an elbow dislocation and to have complained that his knees felt weak, as though ready to dislocate. Two of his brother's 3 children were said to have had multiple dislocations of the patella.

A family study was initiated to document the occurrence of patellar dislocations in the listed relatives and to determine whether other members of the family were affected. The study also was designed to investigate looseness of joints other than the knees and to rule out the various connective tissue disorders in the family.

The criteria used for evaluating joint laxity were those suggested by Carter and Wilkinson (2). In order to be considered loose-jointed, an individual must have 3 of the following: 1) passive apposition of the thumb to the flexor aspect of the forearm, 2) passive hyperextension of the fingers to a position parallel to the forearm, 3) hyperextension of the elbow of more than 190°, 4) hyperextension of the knee of more than 190° and 5) dorsiflexion of the ankle to 45° or less. Based on these criteria, 12 of the 34 examined members of the family had generalized joint laxity (Table 1).

In addition to the 4 members of the family who were reported to have dislocations of the patella, 14 others had hyperextensive knees without actual dislocation. Four individuals with joint laxity of the knees only had children with laxity of other joints. There were 3 cases of congenital hip dislocation in the family. Each case had a parent with laxity of other joints.

DISCUSSION

Generalized joint laxity has been reported to be an autosomal dominant trait with variable expression (3). Carter and Wilkinson (2) reported that there are 2 forms of generalized joint laxity. One type that is transiently present during the neonatal period is limited to females and may have a hormonal etiology. The other form persists into adult life and affects the sexes with equal frequency.

The prevalence of joint laxity in Carter and Wilkinson's survey of 285 English school children was 7%. A similar prevalence was reported by Wynne-Davies (4). However, the latter author pointed out that no children in the first week of life had lax joints and that the period of most loose-jointedness in childhood occurred at about 2 years of age. She also noted a decrease in joint laxity with age. At 6 years of age 5% of her sample were loose-jointed and at 12 years of age only 1% were loose-jointed. Adults normally experience a decrease in joint mobility with increasing age (2).

The wide variation of expression of generalized joint laxity reported by other

TABLE 1. Measurements of Joint Laxity

Individual	Thumb hyper-extension	Finger hyper-extension	Elbow extension >190°	Knee extension >190°	Ankle dorsi-flexion <45°	Congenital hip dis-location
II−1	−	−	−	+	−	
III−1	+	−	−	+	+	
IV−1	+	−	−	−	+	
V−1	+	−	+	−	+	
IV−3	+	−	−	−	+	
V−2	+	+	−	+	+	
V−3	+	−	+	−	+	+
III−2	−	−	−	+	−	
IV−4	+	−	−	−	+	
IV−5	+	−	−	+	+	
V−4	+	+	−	+	−	
V−5	+	+	−	+	−	
IV−6	−	−	+	−	+	
V−6	+	+	+	+	−	
V−7	+	−	−	−	−	
V−8	−	−	+	−	−	
V−9	+	−	−	+	−	
V−10	−	−	−	−	−	
IV−7	−	−	−	+	−	
V−4	−	−	−	−	−	+
V−12	+	+	−	−	−	
IV−8	+	−	+	−	+	
! *V−13*	−	−	−	−	−	
V−14	+	+	−	−	+	+
V−15	+	+	−	+	−	
V−16	+	+	+	+	+	
III−3	−	−	−	+	−	
IV−10	+	−	−	−	−	
V−17	−	−	+	−	−	
V−18	−	−	−	−	−	
III−4	−	−	−	+	+	
IV−12	+	−	+	+	−	
IV−13	+	−	−	+	−	
II−3	+	−	−	−	−	

workers (5) is supported by the observation reported here (Fig. 1 and Table 1). The variation of expression presented a problem in attempting to evaluate the joint laxity objectively. Several members of the family did not meet the criteria for joint laxity proposed by Carter and Wilkinson but had affected children and affected collateral relatives, indicating that they did have the predisposing gene.

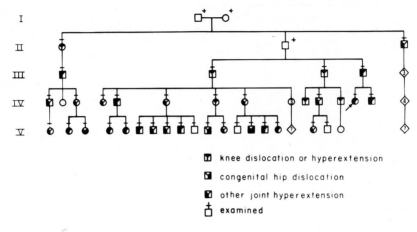

knee dislocation or hyperextension

congenital hip dislocation

other joint hyperextension

examined

Fig. 1.

However, these individuals did have laxity of 1 set of joints in some cases and 2 sets of joints in other cases.

Some cases of joint laxity may result from laxity of the associated ligaments. Brown and Rose (6) recognized this and suggested that isolated laxity of the ligaments (generalized joint laxity) may be the least severe of the connective disorders. The hypermobility syndrome described by Kirk and co-workers (5) may be the consequence of lax ligaments contributing to precocious osteoarthritis and musculoskeletal complaints.

CONCLUSIONS

Joint laxity was investigated in 4 generations of a family. The laxity appeared to be an isolated trait and to be inherited as an autosomal dominant. The knees were the most severely involved joints in affected individuals. In some cases the knees were the only joints involved. These observations support the contention that recurrent knee dislocations, and perhaps congenital hip dislocations, may be a manifestation of generalized joint laxity in affected families.

Based on our studies, we suggest that in some instances it may be appropriate to modify the criteria of Carter and Wilkinson for joint laxity evaluation. In families with documented hypermobility of the joints, it may not be necessary to insist that at least 3 joints be involved. Also, the age of the individual must be considered in such an evaluation.

REFERENCES

1. McKusick, V. A.: "Mendelian Inheritance in Man," Ed. 3. Baltimore: The Johns Hopkins Press, 1971, p. 168.
2. Carter, C. and Wilkinson, J.: Persistent joint laxity and congenital dislocation of the hip. J. Bone Joint Surg. 46B:40–45, 1964.
3. Sturkie, P. D.: Hypermobile joints in all descendants for two generations. J. Hered. 32: 232, 1941.
4. Wynne-Davies, R.: A family study of neonatal and late-diagnosis congenital dislocation of the hip. J. Med. Genet. 7:315–324, 1970.
5. Kirk, J. A., Ansell, B. M. and Bywaters, E. G. L.: The hypermobility syndrome. Ann. Rheum. Dis. 26:417–425, 1967.
6. Brown, A. R. and Rose, B. S.: Familial precocious polyarticular osteo-arthrosis of chondrodysplastic type. N. Z. Med. J. 65:449, 1966.

An X-Linked Form of Cutis Laxa Due to Deficiency of Lysyl Oxidase*

Peter H. Byers, MD,† A. Sampath Narayanan, PhD,
Paul Bornstein, MD, and Judith G. Hall, MD

Cutis laxa is a rare systemic connective tissue disorder which is characterized by marked skin laxity and often by connective tissue abnormalities in other organs (1). Most reported instances are sporadic but several families have been identified in which there is an identifiable pattern of inheritance. In certain families the mode of inheritance is clearly dominant, with individuals of both sexes affected in multiple generations (2–4). In affected members in these families the cutaneous manifestations are usually moderate in severity and the alterations in other organs are minimal. However, in the families in which inheritance appears to be recessive (4, 5), the connective tissue disturbances of organs other than skin may predominate and may contribute to early death.

We have recently identified and studied a family whose members have connective tissue abnormalities consistent with those seen in other forms of cutis laxa but with a mode of inheritance compatible with X-linkage (Fig. 1). Furthermore, we have identified a biochemical disorder in the crosslinking of connective tissue components that appears to account for the clinical findings.

CASE REPORT

Two males *(V-1* and *V-2)*, aged 6 and 9 years when first seen by us, had a constellation of clinical findings which suggested the diagnosis of cutis laxa. Both are similar in appearance with a long thin face, hooked nose, long philtrum, and redundant upper eyelids (Fig. 2). In addition, they have skin which is lax and moderately redundant, especially over the dorsum of the hands and feet (Fig. 3). They have mild joint laxity and skeletal abnormalities consisting of pectus excavatum and carinatum which was severe enough in the older boy to require

*Investigations supported by NIH grants DE-02600, AM-11248 and GM 15253
†Fellow of the Helen Hay Whitney Foundation

Birth Defects: Original Article Series, Volume XII, Number 5, pages 293–298

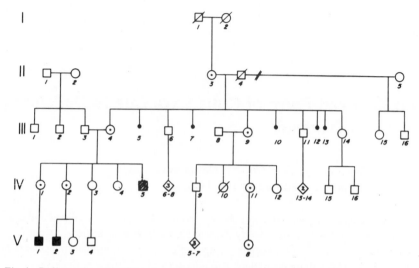

Fig. 1. Pedigree showing affected members and apparent carriers for the X-linked form of cutis laxa; ■ affected males, examined; ▨ presumed affected male, not examined; ⊙ presumed carrier females.

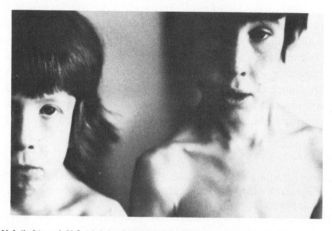

Fig. 2. *V-1* (left) and *V-2* (right), showing the characteristic facies with long thin face, hooked nose, long philtrum, and redundant eyelids. In addition, the axillary wrinkling is apparent.

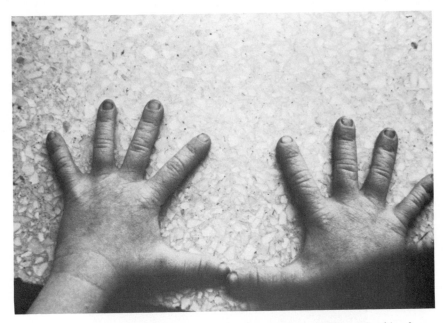

Fig. 3. Dorsum of hands of *V-1* demonstrating the excessive wrinkling of the skin of the fingers.

surgical correction shortly after birth. The major problem which brought both to medical attention was a recurrent urinary tract infection due to multiple large bladder diverticuli resulting in ureteral obstruction and subsequent hydroureter and hydronephrosis (Fig. 4). Although renal function in both is now normal, the older one has an indwelling catheter because of poor bladder function. Both are of normal height, although the older one is the same height as his sister who is one year younger.

The mothers of both boys are sisters *(IV-1* and *IV-2)* and have skin which is softer than usual, with an increase in skin wrinkling, and appear slightly older than their ages. Both had marked joint laxity during childhood which decreased at puberty. A maternal uncle *(IV-5)* died in infancy of pneumonia and was apparently identical in appearance to the 2 affected boys. Two additional female family members are also thought to manifest the affected gene: *IV-11* has increased skin wrinkling and her daughter *(V-8)*, now 2 years old, is markedly hypermobile, as *IV-1* and *IV-2* were as children.

Dermal fibroblasts were obtained and grown from *V-1* and his mother *(IV-1)*. These cells were grown to confluence and the medium harvested and assayed for lysyl oxidase activity using a tritium release assay (6). The results, shown in Table 1, demonstrate that the affected male lacked lysyl oxidase activity using both collagen and elastin substrates, whereas the mother had levels intermediate between those of the affected child and of the control.

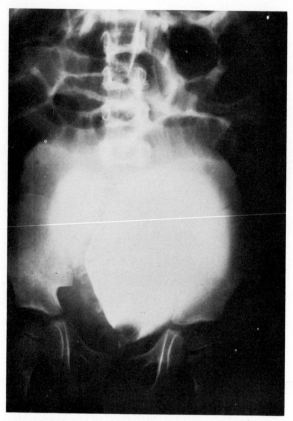

Fig. 4. Voiding cystogram of *V-1* demonstrating the large bladder diverticulum. Other diverticula had been previously removed surgically.

TABLE 1. Lysyl Oxidase Activity in Medium of Cultured Fibroblasts*

| | Tritium Released per mg Protein (cpm) | |
	Elastin substrate	Collagen substrate
CRL 1121 (Control)	4,033	8,771
V-1	320	0
IV-1	3,490	4,752

*Medium from 4 confluent 100 mm cultures of each strain was collected and the proteins precipitated with 40% ammonium sulfate. The precipitate was dissolved with PBS and dialyzed exhaustively against the same buffer. Subsequently, aliquots were incubated at 37° for 4 hours with either chick calvarial collagen or chick aorta elastin radioactively labeled with [4, 5-^{3}H] lysine. The sample was then distilled and the released tritium collected as water and counted in a scintillation counter by standard methods.

These patients have relatively mild cutaneous manifestations of cutis laxa and share some features with patients affected with the Ehlers-Danlos syndrome: they have mild joint hypermobility and form wide scars in addition to having stretchable skin. However, their skin, although stretchable is lax, with wrinkling noticeable especially over the dorsum of the hands and around the axillas (Figs. 2 and 3); they have the characteristic facies of cutis laxa (4), do not bruise easily, do not have fragile skin and skin wounds heal rapidly. Unlike other patients with either cutis laxa or the Ehlers-Danlos syndrome, with the exception of a male with a recessive form of cutis laxa described by Goltz et al (5), their most prominent problem is genitourinary tract diverticulae. These patients differ markedly from the patients recently described by Di Ferrante et al (7) in that they are of normal stature, lack cardiac valvular abnormalities and do not have other forms of congenital heart disease. The latter patients have a connective tissue disorder and also appear to lack lysyl oxidase activity in their dermal fibroblasts. Both the patients described by Di Ferrante et al (7) and those described in this report are quite different from other patients thought to have the X-linked form of the Ehlers-Danlos syndrome previously described by Beighton (8).

The enzyme lysyl oxidase is present in all connective tissues and catalyzes the oxidative deamination of the epsilon amino groups of certain lysyl groups in collagen and elastin and of certain hydroxylysyl groups in collagen. These groups are then involved in condensation with either lysine or hydroxylysine or with their oxidized congeners to form covalent cross-links which are important for the strength of connective tissues (9). Rowe and colleagues (10) have shown that lysyl oxidase is closely linked to the X-chromosomal coat color locus *mottled* in mice. This finding would be consistent with the observations in this report of the absence of lysyl oxidase and the appearance of connective tissue abnormalities in an X-linked pattern of inheritance.

In the presence of the inhibitor of lysyl oxidase, β-aminopropionitrile, there are major connective tissue alterations in almost all organs, although to a variable degree (11). The findings in our patients are clearly much less severe than would be expected if lysyl oxidase activity were completely absent since at present these individuals do not have identifiable cardiovascular or cartilage abnormalities. Thus, it seems possible that there is either tissue-specific control of lysyl oxidase synthesis or, perhaps, collagen type-specific differences in substrate specificity for the enzyme. Possibly, these patients and those described by Di Ferrante (7) have different alleles for a defective lysyl oxidase.

REFERENCES

1. McKusick, V. A.: "Heritable Disorders of Connective Tissues," 4th Ed., St. Louis: C. V. Mosby Co., 1972.
2. Lewis, E.: Cutis laxa. Proc. R. Soc. Med. 41:864, 1948.
3. Sestak, Z.: Ehlers-Danlos syndrome and cutis laxa: An account of families in the Oxford area. Ann. Hum. Genet. 25:313, 1962.

4. Beighton, P.: The dominant and recessive forms of cutis laxa. J. Med. Genet. 9:216, 1972.
5. Goltz, R. W., Hult, A-M., Goldfarb, M. and Gorlin, R. J.: Cutis laxa: A manifestation of generalized elastolysis. Arch. Dermatol. 92:373, 1965.
6. Layman, D. L., Narayanan, A. S. and Martin, G. R.: The production of lysyl oxidase by human fibroblasts in culture. Arch. Biochem. Biophys. 149:97, 1972.
7. Di Ferrante, N., Leachman, R. D., Angelini, P. et al: Lysyl oxidase deficiency in Ehlers-Danlos syndrome type V. Conn. Tiss. Res. 3:49, 1975.
8. Beighton, P.: The Ehlers-Danlos syndrome. London: William Heinemann Medical Books, Ltd., 1970.
9. Gallop, P. M., Blumenfeld, O. O., and Seifter, S.: Structure and metabolism of connective tissue proteins. Ann. Rev. Biochem. 41:617, 1972.
10. Rowe, D. W., McGoodwin, E. B., Martin, G. R. et al: A sex-linked defect in the cross-linking of collagen associated with the mottled locus in mice. J. Exp. Med. 139:180, 1974.
11. Bornstein, P.: The cross-linking of collagen and elastin and its inhibition in osteolathyrism: Is there a relation to the aging process? Am. J. Med. 49:429, 1970.

Generalized Osseous Abnormalities in the Marshall Syndrome*

James J. O'Donnell, MD,† Shari Sirkin, MD and Bryan D. Hall, MD

The Marshall syndrome (1) was originally considered to be a variant of hypo-hidrotic ectodermal dysplasia which was associated with cataracts, sensori-neural deafness and a severe, persistent, flat nasal bridge. It is now clear that this autosomal dominant disorder has little if anything to do with ectodermal abnormalities. Only 4 families (2–4) and 1 isolated case consisting of a total of 17 fully affected individuals have been described in the literature. We wish to expand the spectrum of the Marshall syndrome by presenting an affected father and 2 sons who displayed the characteristic features of that disorder plus generalized osseous abnormalities including thickened calvarium, dural calcifications, beaked or bullet-shaped vertebra in the younger patients, platyspondyly with irregularly concave superior and inferior thoracic and lumbar vertebral margins, small irregular iliac bones, coxa valga, mild radial and ulnar bowing and slightly irregular epiphyses of the limbs. Only the above cranial abnormalities have been noted previously and then in only 2 patients (3, 4); however, as far as we can tell, only a few of the literature cases had general skeletal roentgenograms. We propose that the Marshall syndrome has a generalized osseous component previously unrecognized. We also suggest that the Marshall syndrome is a distinct entity separable from the Stickler syndrome (5–9) and other syndromes involving deafness, eye abnormalities, flat nasal bridge, flat facies and limb abnormalities (10, 11).

*Supported in part by grants from The National Foundation – March of Dimes, The National Institute of General Medical Services (GM-19527), and the Maternal and Child Health Service (MCH 445).

†Recipient of a National Eye Institute Research Fellowship Award (No. EY01054).

Birth Defects: Original Article Series, Volume XII, Number 5, pages 299–314
© 1976 The National Foundation

CASE REPORTS

Case 1

This patient is a 57-year-old white male of German descent. He was originally thought to have congenital syphilis because of a saddle nose and an undocumented but supposedly positive serologic test for syphilis. These findings resulted in his being treated with arsenic injections and mercury rubs when he was 10 years of age. It was also noted during early childhood that he had protruding upper teeth, sensorineural deafness and bilateral cataracts. The cataracts were removed when he was 7 years of age. A rhinoplasty to repair his saddle nose was performed during his midteenage years. Despite the above problems, he experienced good health and was graduated from college with a degree in electrical engineering.

Case 1 is the father of *Cases 2* and *3*, both of whom share similar abnormalities with their father. An additional son and the 30-year-old Yucatan Indian mother of the children are completely normal. The 85-year-old mother of *Case 1* is deaf and has late onset crippling arthritis. Additionally, this woman's brother and sister have early childhood onset of deafness (Fig. 1). The specifics of these historic facts are not available.

Case 1 was examined by us when he was 57 years of age. He was 164 cm tall and had a normal weight and head circumference (Fig. 2A). His downward eye-slant was due to underdeveloped malar bones. The inner canthal (3¼ cm), inter-pupillary (6¼ cm) and outer orbital (10½ cm) distances were normal (Fig. 3A).

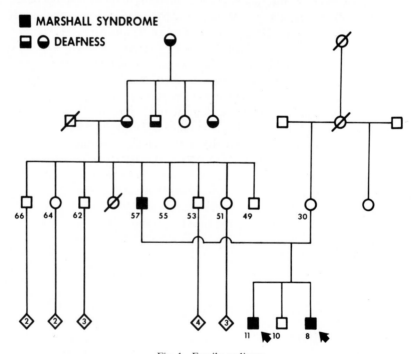

Fig. 1. Family pedigree.

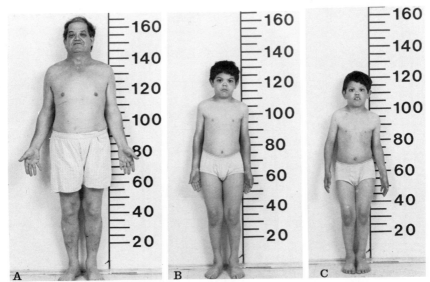

Fig. 2. A) *Case 1*: Note normal proportions and cubitus valgus. B) *Case 2*: Proportional shortened stature with flat facies. C) *Case 3*: Normal proportions, flat facies and shortened stature.

Eye examination showed bilateral aphakia. Pupils, extraocular movements and peripheral visual fields were normal. Corneal diameter was 11.5 mm OU and applanation tension was OD 19 mm Hg and OS 15 mm Hg. There was fibrillar vitreous degeneration. The fundus was normal and specifically did not show radial lattice degeneration. The nose was small and the nasal bridge thin and poorly formed. The philtrum was short and flat (Fig. 4A). There were many dental caries. Examination of the limbs showed an arm span of 152 cm, a 45° cubitus valgus and incomplete elbow extension (Fig. 2A).

Laboratory studies were normal, including BUN, serum creatinine, serum calcium and phosphorus. Serologic studies (FTA-ABS) for syphilis were negative.

Roentgenograms of the entire skeleton revealed a thickened calvarium (Fig. 5A) with extensive calcifications of the falx, tentorium and meninges (Fig. 6); mild midthroracic scoliosis; mild lower thoracic and lumbar platyspondyly with irregular concave surfaces of the superior and inferior lumbar vertebral bodies (Fig. 7); small iliac bones with irregular lateral borders, irregularly fused ischiopubic rami, mild coxa valgus (Fig. 8A); somewhat small and irregular distal femoral and proximal tibial epiphyses (Fig. 9A); bilateral outward bowing of the radius and ulna (Fig. 10A); and wide tufts of the distal phalanges (Fig. 11A).

Case 2

Case 2 is an 11-year-old male. He was the product of a normal term gestation and was delivered by C section because of cephalopelvic disproportion. He weighed 2950 gm and was 48 cm in length. He had a normal neonatal course although possible hearing and visual problems were considered. Bilateral cataracts

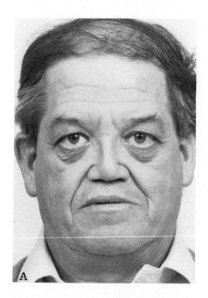

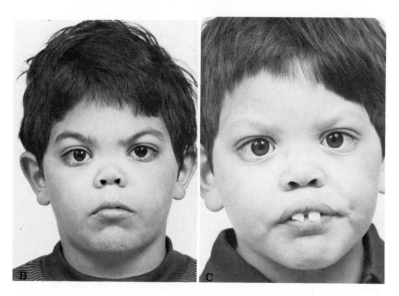

Fig. 3. A) *Case 1*: Flat facies, downward eyeslant and hypoplastic malar bones. B) *Case 2*: Flat nasal bridge, high-set nose and anteverted nares. C) *Case 3*: Note enlarged corneas (right greater than left) and prominent upper central incisors.

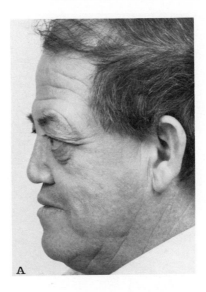

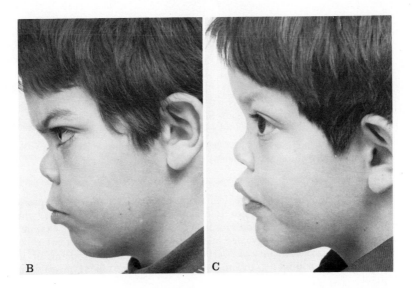

Fig. 4. A) *Case 1*: Flat malar bones and short philtrum. B) *Case 2*: High-set nose, ante-verted nares and prominent inferior border of philtrum. C) *Case 3*: Prominent upper central incisors, prominent inferior border of philtrum, anteverted nares and high-set nose.

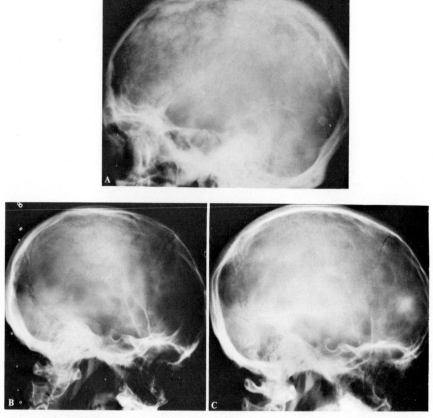

Fig. 5. A) *Case 1*: Thick calvarium with falx, tentorial and meningeal calcifications. B) *Case 2* and C) *Case 3*: Thick posterior calvarium.

were diagnosed at 8 months of age. Clinically, hearing difficulties were present from birth; however, they were not documented until the age of 6 years when he was found to have a 40–45 db sensorineural hearing loss. An eye examination at about the same time revealed bilateral posterior subcapsular cataracts, moderate myopia, visual acuity of 20/200 in each eye and corneal diameter of 12 mm. The boy's general health and psychomotor development have always been good.

Physical examination at 11 years of age showed that the height (125 cm, 10th%), weight (34.4 kg, 50th%) and head circumference (53 cm, 50th%) were within normal limits (Fig. 2B). The midforehead had a prominent fused metopic suture. There was slight asymmetry to the ear placement (right lower than left) and his ears protruded laterally. The malar bones were flat (Fig. 3B). The external eye examination revealed the inner canthal (3½ cm), interpupillary (5¾ cm) and outer orbital (11½ cm) distances to be within normal limits. A right

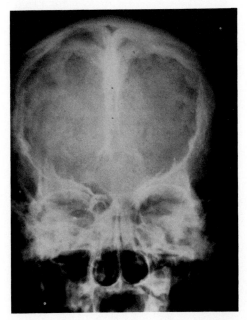

Fig. 6. *Case 1*: Falx, tentorial and meningeal calcification.

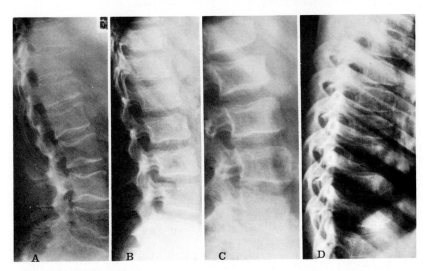

Fig. 7. A) *Case 1*: Platyspondyly with irregular concave inferior and superior vertebral margins. B) *Case 2*: Platyspondyly with irregular vertebral margins. C and D) *Case 3*: C) Platyspondyly with irregular vertebral margins and D) thoracic vertebras showing a bullet shape.

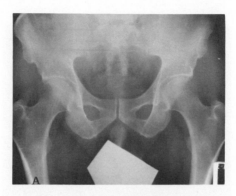

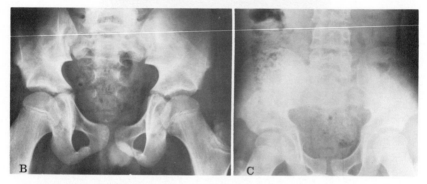

Fig. 8. A) *Case 1*: Small iliac bones with lateral bulges, mild coxa valga and poorly fused ischiopubic rami. B) *Case 2*: Small iliac bones with lateral bulges, mild coxa valga and irregularly fused ischiopubic rami. C) *Case 3*: Similar findings as in *Case 2*.

epicanthal fold was present. Pupils and extraocular movements were normal. Slit-lamp examination showed small posterior subcapsular cataracts. The fundus was a normal myopic one and no radial lattice degeneration of the retina was present. The nose was short and set high upon the face while the nares were upturned and the nasal bridge extremely flat. The philtrum was prominent at its inferior border and the upper central incisors protruded (Fig. 4B). Mild cubitus valgus was present (Fig. 2B).

Laboratory studies were normal, including BUN, serum creatinine, serum calcium, serum phosphorus, FTA-ABS serology, karyotype, urine amino acids and a general urine screen for metabolic disorders including mucopolysaccharides.

A full radiologic survey indicated a thick posterior calvarium (Fig. 5B); mild lower thoracic and lumbar platyspondyly with irregular concave surfaces of the superior and inferior lumbar vertebral bodies (Fig. 7B); mild iliac wing hypoplasia, borderline delayed fusion of the ischiopubic rami, mild coxa valga (Fig. 8B); slightly small proximal tibial epiphyses (Fig. 9B); lateral flattening of the distal tibial epiphyses, relatively long fibulas (Fig. 12A); and small irregular distal radial epiphyses (Fig. 11B).

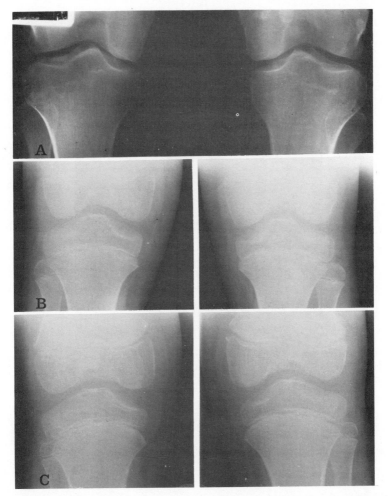

Fig. 9. A) *Case 1*: Small distal femoral and proximal tibial epiphyses. B) *Case 2* and C) *Case 3* show similar findings.

Case 3

This 8½-year-old male was delivered from a normal term pregnancy via a C section with a birthweight of 4970 gm and a birth length of 48 cm. During the neonatal period he was noted to have esotropia and absent nasal bones. He did well in childhood except for recurrent problems of serous otitis media. An audiologic evaluation at 3 years of age elucidated a bilateral 75–85 db sensorineural hearing loss. Following a tonsillectomy-adenoidectomy with bilateral myringotomies, the hearing deficit was reduced to the 50–55 db level. Ophthalmologic examination at 3 years of age showed a 13 diopter myopia, visual acuity of 20/200 and asymmetrically enlarged corneas (right 14 mm, left 13 mm).

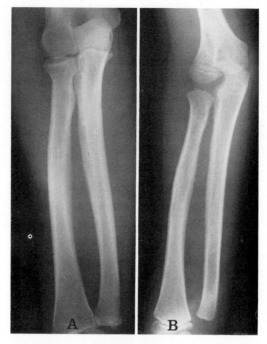

Fig. 10. A) *Case 1* and B) *Case 3*: Slightly bowed radius and ulna with poorly formed radial epiphyses at the wrist.

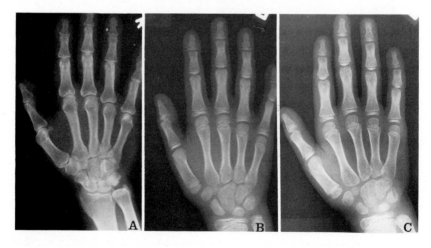

Fig. 11. A) *Case 1*: Wide tufts of the distal phalanges. B) *Case 2* and C) *Case 3*: Poorly formed radial epiphyses at wrist.

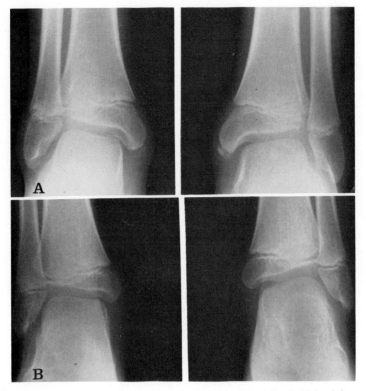

Fig. 12. A) *Case 2* and B) *Case 3* both show lateral flattening of distal tibial epiphyses and relatively long fibulas.

Physical examination at 8½ years revealed a normal height (124 cm, 10th%), weight (28.9 kg, 75th%) and head circumference (Fig. 2C). The skull was normal except for a prominently fused metopic suture. The laterally protruding ears were asymmetrically placed (right lower than left). His malar bones were flat, resulting in a slight downward eyeslant (Fig. 3C). External eye examination showed normal eye measurements for the inner canthal (3¼ cm, 75th%), inter-pupillary (5½ cm, 50th%) and outer orbital (11 cm, 75th%) distances. Vision with correction OD 13.25 was 20/60 and with correction OS 14.00 was 20/70. Pupils and extraocular movements were normal. Corneal diameter OD was 14 mm and OS was 13 mm (normal up to 12 mm). Intraocular pressure under general anesthesia was 8 mm Hg bilaterally. Gonioscopy showed the iris sweeping up to the scleral spur without forming a recess. There was a posterior sub-capsular cataract. The right optic nerve showed the anomalous finding of the inferior nasal vein entering the disk superiorly. The fundus had a normal myopic appearance and there was no radial lattice degeneration of the retina (Fig. 13). His nose was short and set high upon the face while the nares were upturned and the nasal bridge was severely flat. The philtrum was simple,

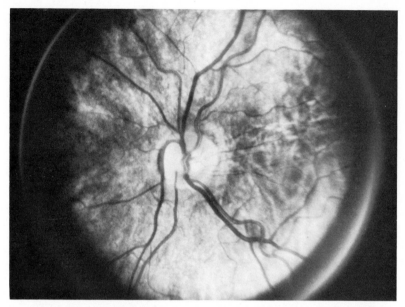

Fig. 13. *Case 3*: Funduscopic picture shows anomalous inferior nasal vein entering the optic disk superiorly and myopic fundus without retinal degeneration or detachment.

short and prominent at its inferior border. The upper central incisors were widely spaced and very prominent (Fig. 4C). A left accessory nipple was present 3 inches below the left nipple. The arm span was 118 cm and a mild cubitus valgus was present (Fig. 2C).

The same laboratory studies that were done in *Case 2* were also negative in *Case 3*. Radiographs showed mild posterior calvarial thickening (Fig. 5C); thoracic and lumbar platyspondyly with bullet-shaped vertebra, some of which have irregular concave superior and inferior margins (Fig. 7C and D); prominent bulges on the lateral iliac bones, mild coxa valga, irregular and unfused ischio-pubic borders (Fig. 8C); small distal femoral and proximal tibial epiphyses (Fig. 9C); flattening of the lateral borders of the distal tibial epiphyses, relatively elongated fibulas (Fig. 12B); slightly bowed radius (Fig. 10B); and poorly formed epiphyses of the radius and ulna at the wrist (Fig. 11C).

DISCUSSION

The family presented herein resembles Marshall's original family in every respect. They resemble the other 3 families subsequently reported by Ruppert et al (2), Keith et al (3) and Zellweger et al (4) except that some of their younger patients had not developed cataracts and a few were not deaf. It appears likely that nearly all patients with the Marshall syndrome eventually develop cataracts, either grossly or by slit-lamp examination.

Our family dramatically illustrates a generalized spondyloepiphyseal disorder which has not been noted in previously reported patients with the Marshall syndrome. The spondyloepiphyseal abnormalities may have been overlooked in previous cases because so very few of those patients had skeletal roentgenograms or because the radiologic findings may be relatively mild during childhood. An indication that radiologic abnormalities existed in previously reported patients can be found in brief observations by Marshall (1) who described calvarial thickening and absent frontal sinuses in the only 2 patients he subjected to craniofacial radiographs. Ruppert et al (2) noted absent frontal sinuses in his adult patient while Keith et al (3) and Zellweger et al (4) noted dural calcifications in their 2 adult patients. The affected father in our family demonstrated falx, tentorial and meningeal calcifications while he and his 2 affected sons all had calvarial thickening. The presence of these relatively unusual radiologic findings in 8 of the 9 patients subjected to craniofacial radiographs, plus the similar clinical features found in all 4 families, strongly suggest that the Marshall syndrome is a distinct entity. If this assumption is correct, then it would be appropriate to theorize that the families reported by Marshall (1), Ruppert et al (2), Keith et al (3) and Zellweger et al (4) will have generalized spondyloepiphyseal abnormalities when subjected to full skeletal radiographs.

The eye changes in our family parallel those in previously reported Marshall syndrome patients. All the patients were highly myopic. All of them have acquired juvenile posterior subcapsular cataracts. It should be noted, however, that one of Marshall's patients probably had congenital cataracts. One of our patients (*Case 3*) had large asymmetric corneas. There had been corneal asymmetry in one (*Case 6*) of Marshall's patients. It is important to distinguish this cause of megalocornea and corneal asymmetry from congenital glaucoma. We have attached no special importance to the unusual optic nerve vein distribution in *Case 3* (Fig. 13). There is high myopia but no retinal degeneration or detachment in our family. Two retinal detachments have been reported in the Marshall syndrome, but these were presumably secondary retinal detachments occurring after the lens was removed by cataract surgery (Ruppert et al (2), Case 2) and by traumatic dislocation (Marshall (1), Case 3). This may be one of the distinguishing features when comparing the Marshall and Stickler syndromes as the latter has a high incidence of spontaneous juvenile retinal detachment with the lens in place.

The Marshall syndrome (1—4) appears to be a clinically consistent syndrome. This is at great variance to the Stickler syndrome (5—7, 9) which is an autosomal dominant disorder with an extreme degree of clinical variability (7). Table 1 compares the similarities and differences of these syndromes.

Cohen (16) has suggested that the Marshall and Stickler syndromes are one and the same. There is no doubt that the spondyloepiphyseal findings in our

TABLE 1. Clinical Comparison of the Marshall and Stickler Syndromes

	Marshall Syndrome (20 patients) (Present article plus references 1–4)			Stickler Syndrome (33 patients) (References 5–7, 9, 12–15)*		
Craniofacial Features	Affected	Examined†	% Affected	% Affected	Affected	Examined†
Mental retardation	5	20	**25**	12	3	33
Deafness	16	20	**80**	18	6	33
Myopia	19	19	100	83	24	29
Cataract	12	20	**60**	34	11	32
Retinal detachment	2	20	10	**48**	15	31
Flat nasal bridge	20	20	**100**	29	9	31
Anteverted nostrils	20	20	**100**	0	0	21
Flat malar bones	20	20	**100**	41	9	22
Cleft palate, isolated	1	19	5	**31**	10	32
Pierre Robin anomaly	0	20	0	**16**	5	32
Prominent upper incisors	8	12	67	0	0	32
Limb Features						
Hyperextensible joints	0	20	0	**48**	13	27
Hypoextensible joints	4	5	**80**	17	5	29
Arthropathy‡	0	20	0	**33**	11	33
Thin habitus	0	20	0	**21**	7	33
Shortened stature§	10	12	**83**	6	2	33

*The family of Schreiner et al (8) is not used to compute this table because of mixed contributions from both sides.

†Examined means those patients in which a judgment (historic or clinical) could be made.

‡Arthropathy means swollen tender joints with or without arthritis.

§Shortened stature means 10th percentile or below.

Bold face numbers signify major differences in the 2 syndromes.

family and those of some of the Stickler diagnosed families are similar; however, not one of the patients diagnosed as having the Stickler syndrome have had thickened calvarium or falx, tentorial or meningeal calcifications. Clinically, every patient with the Marshall syndrome has had congenital and persistently severe flat nasal bridge, nostril anteversion and malar hypoplasia. In only one patient with the Marshall syndrome has there been a cleft palate while approximately 47% (Table 1) of the Stickler syndrome patients have had cleft palates with or without the Pierre Robin anomaly. Deafness is also much less frequent in the Stickler syndrome and it is usually conductive in nature (7) while Marshall syndrome patients usually have sensorineural deafness (4). There are many other differences between these 2 disorders and the reader is referred to the table for more detail.

We have presented a family with the Marshall syndrome. They adhere clinically and genetically to the previously reported cases. The Marshall syndrome is a distinct entity separable from other similar entities, particularly the Stickler syndrome. It appears clear that the Marshall syndrome is associated with craniofacial abnormalities and generalized spondyloepiphyseal abnormalities.

REFERENCES

1. Marshall, D.: Ectodermal dysplasia: Report of a kindred with ocular abnormalities and hearing defect. Am. J. Ophthalmol. 45:143, 1958.
2. Ruppert, E. S., Buerk, E. and Pfordresher, M. F.: Hereditary hearing loss with saddle-nose and myopia. Arch. Otolaryngol. 92:95, 1970.
3. Keith, C. G., Dobbs, R. H., Shaw, D. G. and Cottrall, K.: Abnormal facies, myopia, and short stature. Arch. Dis. Child. 47:787, 1972.
4. Zellweger, H., Smith, J. K. and Grutzner, P.: The Marshall syndrome: Report of a new family. J. Pediatr. 84:868, 1974.
5. Stickler, G. B., Belau, P. G., Farrell, F. J. et al: Hereditary progressive arthroopthalmopathy. Mayo Clin. Proc. 40:433, 1965.
6. Stickler, G. B. and Pugh, D. G.: Hereditary progressive arthro-ophthalmopathy II. Additional observations on vertebral abnormalities, a hearing defect, and a report of a similar case. Mayo Clin. Proc. 42:495, 1967.
7. Opitz, J. M., France, T., Herrmann, J. and Spranger, J. W.: The Stickler syndrome. N. Engl. J. Med. 286:546, 1972.
8. Schreiner, R. L., McAlister, W. H., Marshall, R. E. and Shearer, W. T.: Stickler syndrome in a pedigree of Pierre Robin syndrome. Am. J. Dis. Child. 126:86, 1973.
9. Hall, J.: Stickler syndrome: Presenting as a syndrome of cleft palate, myopia and blindness inherited as a dominant trait. In Bergsma, D. (ed.): "Clinical Cytogenetics and Genetics," Birth Defects: Orig. Art. Ser., vol. X, no. 8. Miami: Symposia Specialists for The National Foundation—March of Dimes, 1974, pp. 157–171.
10. Wagner, H.: Ein bisher unbekanntes Enbleiden des Auges (Degeneratio hyaloideoretinalis hereditaria) beobachtet im Kanton. Zurich. Klin. Monatsbl. Augenheilkd. 100:840, 1938.

11. Cohen, M. M., Jr., Knobloch, W. H. and Gorlin, R. J.: A dominantly inherited syndrome of hyaloideoretinal degeneration, cleft palate and maxillary hypoplasia (Červenka syndrome). In Bergsma, D. (ed.): Part XI. "Orofacial Structures," Birth Defects: Orig. Art. Ser., vol. VII, no. 7. Baltimore: Williams & Wilkins Co., for The National Foundation—March of Dimes, 1971, pp. 83–86.

12. Walker, B. A.: A syndrome of nerve deafness, eye anomalies and marfanoid habitus with autosomal inheritance. In Bergsma, D. (ed.): Part XI. "Ear," Birth Defects: Orig. Art. Ser., vol. VII, no. 4. Baltimore: Williams & Wilkins Co., for The National Foundation—March of Dimes, 1971, pp. 137–139.

13. Schimke, R. N.: Pierre Robin syndrome in sibs. In Bergsma, D. (ed.): Part II. "Malformation Syndromes," Birth Defects: Orig. Art. Ser., vol. V, no. 2. White Plains: The National Foundation—March of Dimes, 1969, p. 222.

14. Smith, W. K.: Pierre-Robin Syndrome in brothers. In "Malformation Syndromes," Ibid., p. 220.

15. Opitz, J. M.: Ocular anomalies in malformation syndromes. Trans. Am. Acad. Ophthalmol. Otolaryngol. 76:1193, 1972.

16. Cohen, M. M.: The demise of the Marshall syndrome. J. Pediatr. 85:878, 1974.

Deafness and Vitiligo

Theodore F. Thurmon, MD,* Jennifer Jackson, BS and Cynthia G. Fowler, MA

An isolated Louisiana village has been the site of genetic investigations over a period of years. Rosenbaum et al (1) reported on diabetic and prediabetic nephropathy in a family from this village (their Cases 8, 9, 10, 12 and 13). Kloepfer et al reported on inherited microcephaly in this village (2). During our investigation of the manifestations of the founder effect in various types of populations (3), we encountered a kindred in which autosomal recessive genes for congenital deafness and vitiligo appeared to be segregating.

The summary family pedigree (Fig. 1) indicates that there were 5 children with congenital deafness in 1 sibship, 3 of whom also had vitiligo. Two of their male first cousins had only congenital deafness and 2 more remote cousins had only vitiligo. Two of their ancestors in generation IV had congenital deafness. In this pedigree all persons with either congenital deafness or vitiligo had consanguineous parents and all were descended from the same ancestral couple.

All of the deaf children were noticed in the first few months of life to lack hearing. Audiograms were strikingly similar, each showing severe bilateral nonprogressive sensorineural hearing loss (Fig. 2A).

Audiograms were also done on the persons affected by vitiligo alone – the parents of one of the sibships with deafness and the parents of one of the persons with vitiligo. Subject *VIII-1*, affected by vitiligo, had high frequency hearing loss of the type which is characteristic of noise-induced hearing loss (Fig. 2B). The same was found in *VII-10*, father of a sibship with deafness, and in *VIII-17*, father of a child with vitiligo. Subject *VII-11*, mother of the sibship with deafness, had a generalized slight flat curve hearing loss (Fig. 2C).

*Supported by The National Foundation–March of Dimes grant CE-39

Birth Defects: Original Article Series, Volume XII, Number 5, pages 315–320
© **1976 The National Foundation**

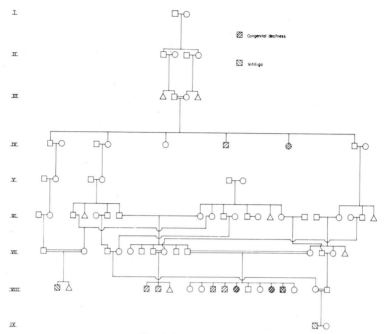

Congenital deafness

Vitiligo

Fig. 1. Summary pedigree.

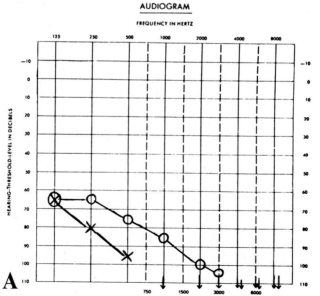

Fig. 2. Audiograms. A) *VIII-14*, showing profound loss. B) *VIII-1*, showing abrupt high frequency loss. C) *VII-11*, showing mild flat loss.

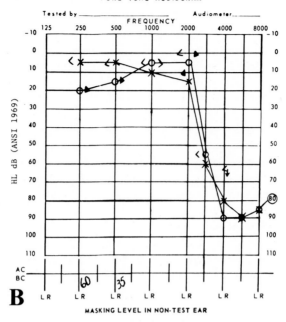

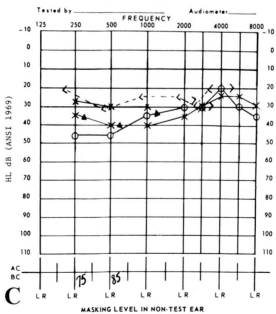

Fig. 2 (Continued)

The parents of each subject with vitiligo gave similar histories. They first noticed the children to have small depigmented areas around the waist and in the flexion creases on the limbs when the children were from 6 to 8 years of age. The areas slowly enlarged and other areas became involved in a similar manner. No area of the body was spared, though there was much individual variation in the involvement.

Neither the deafness nor the vitiligo was different in the subjects with both disorders.

The village has several interesting aspects from the genetic standpoint. It was formed during the latter part of the 18th century and was bounded by Acadian, Spanish and German settlement areas. It was a small area of high land totally surrounded by nearly impenetrable swamp. Though bounded by other settlement groups, the village was remote from them, and the early settlers consisted of persons who for one reason or another wished to remove themselves from one of the other settling groups. Most of these persons were outcasts for either social or legal reasons. During the formation of the settlement, there was little marriage between relatives. Shortly thereafter, however, there began a long period of endogamy which has continued until the present. The nature of the original settlers caused them to be shunned by the surrounding groups. Another factor in their isolation was local folklore about the amount of unusual disease in the offspring of these people. Though the first reason has been lost to history, the fear of defective offspring has persisted to the present. Parents from the surrounding communities frown upon social relationships between their children and children from this isolated village. Marriage between people from the village and those from outside communities is still uncommon. The villagers themselves are keenly aware of these prejudices and point with some pride at the occasional marriages between villagers and outsiders as signs of progress. There are now about 500 inhabitants of the village.

DISCUSSION

Autosomal recessive inheritance of congenital deafness is well established (4). Fraser estimated that nearly half of congenital deafness was of this variety (5) and Morton estimated the proportion at greater than half (6). Inherited deafness in association with various pigmentary disorders has also been previously described (5, 7). To our knowledge there are no reports of congenital deafness associated with vitiligo, although Reed et al (7) suggested that such might exist. In the albinism-deafness syndrome, electron microscopy demonstrates that the melanocytes in the congenitally hypomelanotic areas of skin have been replaced by Langerhans cells. Vitiligo appears to be an acquired rather than congenital defect but is characterized by the same finding on electron microscopy.

The exact relationship between the vitiligo and the deafness in this kindred is not clear; however, the data support several possibilities. All the cases of vitiligo in the village are descended from a single ancestral couple and all cases have consanguineous parents. There are no instances in which a parent and child in the same household or an unrelated person in the same household has vitiligo. These data appear to favor autosomal recessive inheritance rather than other forms of inheritance, or an infectious agent, as the etiology of the vitiligo in this kindred. The gene, if any, for vitiligo appears to be unlinked to, and separate from, that for the congenital deafness. Those children who have both conditions showed no ameliorating or potentiating effect of either condition on the other.

The audiometric findings in the parents of the deaf children, and in the children affected with only vitiligo and their parents, are remarkable in the number of abnormalities found. Few published data are available on audiometric findings in parents of deaf children, and no consistent abnormalities have been noted. Particularly impressive in this kindred was the number of persons having high frequency noise-induced hearing loss. Each of these persons had had exposure to noises which could account for the loss. The tendency to acquire noise-induced hearing loss is known to be variable. The data from this pedigree suggest the possibility that heterozygosity from a gene for autosomal recessive congenital deafness may predispose the individual to noise-induced hearing loss.

Great caution must be used in interpreting data from inbred groups. Cases of diseases of low prevalence found by chance in these groups may well have consanguineous parents. If the group is quite small, many persons may also have descended from one ancestral couple. In spite of these reservations, the value of such groups in establishing autosomal recessive inheritance of diseases is well known. This would appear to be the first report of data supporting autosomal recessive inheritance of vitiligo.

Because of the microscopic similarity of the 2 diseases of piebaldness and vitiligo, it would be possible to confuse them if data on age of onset were not available. For this reason, caution is indicated when interpreting familial findings of piebaldness and deafness (8). It is possible that some previous cases of this condition in small inbred groups may represent chance associations of vitiligo and deafness as demonstrated in this pedigree.

SUMMARY

Seven cases of congenital deafness and an autosomal recessive pedigree pattern were observed in an inbred kindred. Three of the cases also had vitiligo, and there were 2 other cases of vitiligo in this kindred. The vitiligo also appeared to have an autosomal recessive inheritance pattern.

ACKNOWLEDGMENTS

Dr. Jose E. Torres referred the family to us. Deborah J. Jessup, B.S. assisted in the family studies. Dr. Donald L. Rampp, Head of the Department of Audiology and Speech Pathology, School of Allied Health Professions, and Mr. Elton A. Kamkin, Superintendent of the Louisiana State School for the Deaf, cooperated in the study.

REFERENCES

1. Rosenbaum, P., Kattine, A. A. and Gotsegen, W. L.: Diabetic and prediabetic nephropathy in childhood. Am. J. Dis. Child. 106:83-95, 1963.
2. Kloepfer, H. W., Platou, R. J. and Hansche, W. J.: Manifestations of a recessive gene for microcephaly in a population isolate. J. Genet. Hum. 13:52-59, 1964.
3. Thurmon, T. F. and DeFraites, E. B.: Effect of size of founding group on founder effect. (Abstract) Am. J. Hum. Genet. 25:80A, 1973.
4. McKusick, V. A.: "Mendelian Inheritance in Man," 4th Ed. Baltimore: The Johns Hopkins Press, 1975, pp. 398-399.
5. Fraser, G. R.: Profound childhood deafness. J. Med. Genet. 1:118-151, 1964.
6. Morton, N. E.: The mutational load due to detrimental genes in man. Am. J. Hum. Genet. 12:348-364, 1960.
7. Reed, W. B., Stone, V. W., Boder, E. and Ziprkowski, L.: Pigmentary disorders in association with congenital deafness. Arch. Derm. 95:176-186, 1967.
8. Woolf, C. M., Dolowitz, D. A. and Aldous, H. E.: Congenital deafness associated with piebaldness. Arch. Otolaryng. 82:244-250, 1965.

Tumoral Calcinosis and Engelmann Disease

Theodore F. Thurmon, MD* and Jennifer Jackson, BS

The instructions from the Bureau of the Census to its enumerators for the 1950 census of the United States included an order to "report persons of mixed white, Negro, and Indian ancestry living in the eastern United States in terms of the name by which they are locally known". (1). In a summary of the data from this census, Beale (2) included under the name "other" a small group of people in St. Landry Parish, Louisiana. In a later summary Witkop (3) termed this group the St. Landry Mulattoes. The origin of the group is obscure, but they are well known locally and are the subject of much folklore pertaining to prejudices against miscegenation. Many prominent local families know, or fear, that some of their ancestors participated in the founding of the group. Basically the group constitutes persons of varying degrees of admixture among Amerindian, black and white stock. They have been socially isolated and inbreeding has become common. There is much local medical folklore about unusual diseases in this group thought to result both from the inbreeding and the miscegenation, so we have begun a small-scale investigation of them. During this investigation we have encountered a family in which the diseases of tumoral calcinosis and Engelmann disease appear to be segregating. A third condition, snow-capped teeth, was also found, but it is not clear whether this is a separate entity or a previously unrecognized effect of Engelmann disease.

CASE REPORTS

The summary family pedigree (Fig. 1) illustrates that among the sibs in the last generation, one has tumoral calcinosis, one has Engelmann disease and one

*Supported by The National Foundation—March of Dimes grant CE39

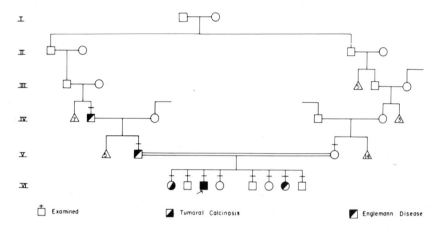

Fig. 1. Summary pedigree.

has both of the disorders. The parents are consanguineous, and the father has Engelmann disease as does the paternal grandfather.

Subjects *IV–8* and *V–6*, who had Engelmann disease, were asymptomatic but had noticeably enlarged bones in their forearms. Subjects *VI–3* and *VI–7* had milder but similar findings in the forearms. They had complained of evanescent pains and aches in the arms and legs since early childhood. Aspirin significantly relieved the pains. Their radiographs as well as those of *V–6* (Fig. 2), showed symmetric bilateral cortical thickening of the diaphyses, typical of Engelmann disease. Subjects *VI–1* and *VI–3* had developed painful firm tumoral masses several times after local trauma. Neither had angioid streaks

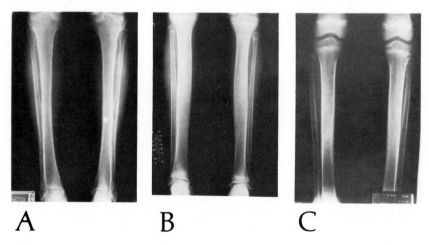

Fig. 2. Radiographs of forelegs showing symmetric bilateral cortical thickening of the diaphyses. A) *V–6*, B) *VI–3* and C) *VI–7*.

of the retina. Radiographs (Fig. 3) showed the masses to have rounded lobu-
lated calcification typical of tumoral calcinosis. Surgery in each instance resulted in successful removal of the masses. There had been no recurrences while
the subjects were on a low phosphorus diet. Subject *VI−1* had no limb pains
other than those associated with the tumors. Her radiographs were equivocal,
some areas suggesting Engelmann disease (see clavicles, Fig. 3A). Her blood
phosphorus was elevated to 6.3 mg% (normal range 2.5−4.5 mg%), as is seen
with Engelmann disease. The blood phosphorus was also elevated in subjects
VI−3 (6.6 mg%) and *VI−7* (5.7 mg%).

Subjects *V−6, VI−1, VI−3* and *VI−7* all had snow-capped teeth (4). The
tips of the teeth from the incisors to the 3rd molars had a ground glass appearance, probably representing a type of local hypoplastic amelogenesis
imperfecta.

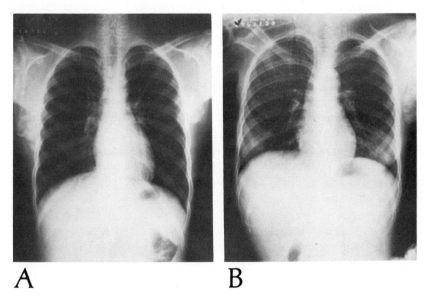

A   B

Fig. 3. Radiographs of thorax showing typical masses of tumoral calcinosis adjacent to
scapula. A) *VI−1* and B) *VI−3.*

DISCUSSION

The presence of tumoral calcinosis in this family attests to the African genes
present in the group to which they belong. With the exception of 3 cases, all
reports of tumoral calcinosis have been in blacks. Two of the reports of tumoral
calcinosis in whites (5, 6) were from North Carolina, an area in which triracial
groups similar to the St. Landry group are found, and the other report (7) was
of a Brazilian from an area also characterized by triracial groups. Though none
of those reports contained any treatment of the subject of ethnic origin, it is

reasonable to suspect that racial admixture may have occurred for each of those cases. Though the disorder of tumoral calcinosis is inherited as an autosomal recessive trait, local trauma is probably necessary to cause its clinical manifestation. Most cases reported have been found to have elevated blood phosphorus (8), so it may be possible to identify sibs who have not yet manifested the disorder. It is of interest that in the sibship reported here, the child shown as having only Engelmann disease did have an elevated blood phosphorus.

Sclerotic bone diseases have many similarities and so have been confused in reports in the literature. Some cases of Engelmann disease may have a characteristic symptomatology associated with them; however, other cases are diagnosed simply by the distribution of the roentgenographic bone findings and on the absence of characteristics of other disorders with similar roentgenographic findings. Consequently, it is reasonable to treat a diagnosis of Engelmann disease with reservations. None of the cases reported here had the characteristic thin limbs, short stature, muscle atrophy and weakness which had been reported in some cases of the disease (9). Great variability of the phenotype of Engelmann disease is well documented (10), only one third of patients having any of these findings. The fact that none of the patients had any signs of other diseases which have similar roentgenographic findings further strengthens the diagnosis in this family. It is notable that of the 20 families reported in the literature, 3 had none of the findings listed above, and the diagnosis was made by roentgenographic findings only. This may mean that there is heterogeneity in the diagnostic category of Engelmann disease, the family reported here representing a mild type.

In the few cases of snow-capped teeth which have been reported, no mention has been made of any associated disorders (4), but evidently familial cases have been seen in an apparent autosomal dominant pattern. In the family reported here, the dental disorder was found in all those with Engelmann disease except *IV–8*. Subject *IV–8* had lost most of his teeth and could not recall any unusual characteristics of the teeth. It is quite possible that the dental abnormality is a part of Engelmann disease, at least the type of Engelmann disease seen in this family, but it is also possible that the dental abnormality is a separate autosomal dominant condition and the association is due to chance and a small sample size.

SUMMARY

Four cases of Engelmann disease were observed in a dominant inheritance pattern in a family of St. Landry Mulattoes. In the youngest sibship, one of the cases also had tumoral calcinosis. Another sib had only tumoral calcinosis, though some findings suggested that this sib may also have had Engelmann disease. All 3 of the younger sibs had snow-capped teeth. It is suggested by this family that snow-capped teeth may be part of the syndrome of Engelmann disease.

ACKNOWLEDGMENTS

Dr. Edward S. Homan referred the family to us. Deborah J. Jessup, B. S. assisted in the pedigree studies. Dr. Dan Voorhies, Director of Lafayette Charity Hospital, Dr. Tipton McKnight, Director of Earl K. Long Memorial Hospital, and Dr. Harry Dascomb, Director of Charity Hospital in New Orleans, cooperated in the study.

REFERENCES

1. "Enumerators Reference Manual, 1950 Census of the United States." U. S. Dept. Comm. Washington, D. C., p. 34.
2. Beale, C. L.: American triracial isolates. Their status and pertinence to genetic research. Eugenics Quart. 4:187–196, 1957.
3. Witkop, C. J., MacLean, C. J., Schmidt, P. J. and Henry, J. L: Medical and dental findings in the Brandywine isolate. Ala. J. Med. Sci. 3:382–403, 1966.
4. Witkop, C. J. and Rao, S.: Inherited defects in tooth structure. In Bergsma, D. (ed.): Part XI. "Orofacial Structures," Birth Defects: Orig. Art. Ser., vol. VII, no. 7. Baltimore: Williams and Wilkins Co. for The National Foundation—March of Dimes, 1971, pp. 153–184.
5. Barton, D. L. and Reeves, R. J.: Tumoral calcinosis: Report of three cases, and review of the literature. Am. J. Roentgen. 86:351–358, 1961.
6. McPhaul, J. J. and Engel, F. L.: Heterotopic calcification, hyperphosphatemia, and angioid streaks of the retina. Am. J. Med. 31:488–492, 1961.
7. Lafferty, F. W. and Pearson, O. H.: Tumoral calcinosis. Am. J. Med. 38:105–118, 1965.
8. Lafferty, F. W. and Pearson, O. H.: Skeletal, intestinal, and renal calcium dynamics in hyperparathyroidism. J. Clin. Endoc. Metab. 23:891–902, 1963.
9. Holmes, L. B., Moser, H. W., Halldorsson, S. et al: "Mental Retardation. Atlas of Diseases with Associated Physical Abnormalities." New York: Macmillan, 1972, p. 262.
10. Sparkes, R. S. and Graham, C. B.: Camurati-Engelmann disease. Genetics and clinical manifestations with a review of the literature. J. Med. Genet. 9:73–85, 1972.

Bilateral Hammertoes: A New Dominant Disorder

William B. Reed, MD **and Les Schlesinger,** MD

The proband is a 70-year-old retired Italian-American who complained of a chronic maceration under both his 2nd and 3rd toes. Upon examination he was found to have had hammertoes of the 2nd toes accounting for his symptoms (Fig. 1). The toe deformity had been present since birth. His mother and half brother also had this deformity. His mother's first marriage produced 4 boys and 1 girl, with the patient the only one affected. He had no children. His mother's second marriage yielded 3 children (2 males, 1 female); one of the male sibs was affected. None of the affected individuals have other skeletal dysplasias.

A search of the literature reveals no such autosomal dominant skeletal dysplasia. Victor McKusick believes that this could be a new entity (personal communication, June 1975).

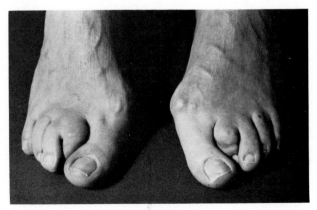

Fig. 1. Note the bilateral hammertoe deformity with soft tissue webbing of the 2nd and 3rd toes.

Selected Abstracts

AN APPROACH TO THE DIAGNOSIS OF SYNDROMES OF UNKNOWN ETIOLOGY

M. Preus and F. C. Fraser

MRC Genetics Group, The Montreal Children's Hospital and Dept. of Biology, McGill University, Montreal, Canada

The concept of a syndrome implies that if a sufficient number of anomalous features occur together a sufficient number of times (though no one feature need always be present), then their association reflects an underlying common cause. When the cause is not known, it becomes a problem to decide how many of which features need to be present in a subject to justify a diagnosis of the syndrome. Diagnostic indexes, such as the Walker dermatoglyphic index for Down syndrome, are based on the relative frequencies of various features in the syndrome vs the general population. When the cause is not known, one must resort to estimating the frequencies of features from cases diagnosed on the basis of a priori assumptions about which features characterize the syndrome. However, if a group of patients suspected of having a syndrome in fact consists of a mixture of bona fide cases and cases representing coincidental associations of certain features, and if the index includes features which were not of major importance in ascertainment, the distribution of indexes for the group should be bimodal. This was demonstrated for a group of patients referred because of suspected Down syndrome, where the karyotype subsequently provided a definitive diagnosis, thus demonstrating the validity of the approach. Similar bimodality appears in the distribution of an index developed for the de Lange syndrome (in which the cause is not known), suggesting that the diagnostic index can identify a major portion of suspected cases as genuine.

FAMILIAL OCULODENTODIGITAL DYSPLASIA

M. Waziri, H. Zellweger and V. Ionasescu

Dept. of Pediatrics, University Hospitals, Iowa City, Iowa

A family including a 35-year-old mother and 2 of her 4 children (12-year-old boy, 11-year-old girl) with oculodentodigital dysplasia is reported. The clinical findings show sunken eyes, epicanthic folds, microcornea, long thin nose

Birth Defects: Original Article Series, Volume XII, Number 5, pages 329–333

without alar flare, low-set cup-shaped ears, hypoplasia of the enamel, campto-dactyly of the 5th finger and syndactyly of the 4th and 5th fingers. The older child also shows left congenital hip dislocation, hypoplasia of middle phalanx of the 5th finger, mild mental retardation (full scale IQ of 70), corticospinal tract involvement, and coronal hypospadias. The younger child presents in addition partial syndactyly between 2nd and 3rd toes and convergent squint. Her intelligence is normal. Chromosomal studies, evaluation of plasma and urinary amino acids are normal.

PSEUDO-RUBINSTEIN SYNDROME: AN AUTOSOMAL DOMINANT CONDITION

Y. Lacassie

Division of Medical Genetics, Dept. of Medicine, Johns Hopkins University School of Medicine, Baltimore, Maryland

At the First Conference on the Clinical Delineation of Birth Defects, Dr. Meinhard Robinow described a mother and all her 4 children with a disorder resembling the Rubinstein-Taybi syndrome (Part II. Malformation Syndromes, Birth Defects: Orig. Art. Ser. $V(2)$:42, 1969). Tall stature, less severe mental deficiency, absence of cryptorchidism or associated physical abnormalities and unremarkable dermatoglyphics in these patients made him suggest that they represented a "new" syndrome. The following case report gives support to Dr. Robinow's communication and will help delineate this new entity further.

The patient was referred to the Division of Endocrinology and Genetics, Hospital M. Arriarán (Santiago, Chile), when he was 13 months old because of delayed milestones and abnormal spatulate fingers. The referring diagnosis was probable Rubinstein-Taybi syndrome. The propositus was the first son of an apparently healthy 30-year-old father and a 22-year-old unrelated mother. She had broad thumbs as had her 18-year-old brother who was also mentally retarded. The patient was the product of a full-term undesired pregnancy, the mother having received 2 unsuccessful shots (Tocafinal) with abortive intention 6 weeks into her pregnancy. Delivery was on 8/5/71, with forceps because of fetal stress. Birthweight 3520 gm and height 51.5 cm. Placenta was normal (640 gm). Slight hypoxia and Apgar 7 at the first minute were recorded. Psychomotor development was slow and because of this the mother consulted a neurologist at the referring hospital. At 13 months the patient was a hyperactive and taller than average infant (80 cm) with a head circumference of 48 cm. The face was without evident dysmorphism, but apparent hypertelorism, well-developed nasal bones and rather small mandible gave a slight triangular-shaped facial appearance. His visual acuity was poor as he was unable to follow objects, and the fundi showed slight peripheral salt-and-pepper pigmentations and a dull fovea. Normal external male genitals were present. Limbs were slightly hypertonic. Broad thumbs and slight spatulate fingers were noted and both toes were broader than usual. Dermatoglyphics revealed 4 whorls and 6 ulnar loops on the fingers with a high total ridge-count (224), intermediate axial triradius and transitional II simian crease on the right hand. Distal loops were present on the 4th interdigital areas and there were no patterns on the thenar or hypothenar areas. EEG and urine chromatography were normal. Radiographs showed short and wide terminal phalanges on thumbs and great toes. No abnormalities were found on the spine or pelvis. VDRL and toxoplasmosis tests were normal on both the mother and son.

In the mother, broad thumbs and borderline wide great toes were present. Dermatoglyphics were similar to those on her son. No other physical abnormalities were detected and mental development seemed to be within the normal range. We had the opportunity to see an affected maternal uncle whose broad thumbs and great toes were striking. Mild mental retardation was evident. No other obvious physical abnormalities were noted.

At 2 years of age psychomotor development had progressed very slowly and an autoagressive trend was present. Blood uric acid levels were normal; the fundi had not changed; and the rate of growth was normal.

This family and the original one of Dr. Robinow, as well as 2 similar families I have known, one of which was personally studied 8 years ago, suggest an autosomal dominant pattern of inheritance with variable expressivity for this syndrome.

CRANIOCARPOTARSAL DYSPLASIA WITH PARTIAL MANDIBULAR AGENESIS — AN EMBRYOLOGIC RARITY OR UNUSUAL GENE EXPRESSION?

R. G. Sanger

University of Southern California School of Dentistry and Children's Hospital of Los Angeles, Los Angeles, California

Since Freeman and Sheldon first described craniocarpotarsal dysplasia in 1938, 40 or more additional cases have been described in the literature. Sporadic cases, as well as those showing autosomal dominant inheritance with variable expressivity, have been reported which postulate that certain genetic factors might encourage mesodermal disorganization and subsequent dysmorphogenesis in the 3 major skeletal areas. In one of these areas, the face, facial and palatopharyngeal muscular dystrophy and paralysis have been observed. Pertinent facial features have included: ocular hypertelerism, small nose with bent alae simulating colobomas of the nostrils, long philtrum, microstomia and protruding or whistling lips. A recent review of all cases noted a characteristic groove (H-shaped or V-shaped) of the chin. Also, some authors have reported bilateral mandibular alveolar grooves or incomplete clefts.

I have examined a 3-month-old infant with the classic features of craniocarpotarsal dysplasia with an unusual chin and mandibular body defect. Intraoral exam revealed the mandible to have a complete bony cleft at the symphysis. However, AP and lateral cephalograms revealed: 1) the mandible to be structurally deficient anterior to the second deciduous molars bilaterally and 2) a small bony remnant located in the midline and superior to the hyoid. Postmortem intra-oral dissection revealed this mandibular deficiency and the muscles that would normally attach at the genial-symphyseal area of the mandible were traced to the bony remnant. Furthermore, the mental foramen was dislocated posteriorly on the mandibular body. Therefore, this does not represent a central or symphyseal cleft of the mandible, but rather total bilateral clefting or lack of structural development in the region where other authors have noted alveolar grooves and/or partial clefts with a chin defect. This possibly represents the most unusual expression of this dysplasia in the facial region.

THE SYNDROME OF APLASIA CUTIS CONGENITA WITH TERMINAL TRANSVERSE DEFECTS OF LIMBS

N. Scribanu and S. A. Temtamy

Dept. of Pediatrics, Georgetown University Medical Center, Washington, D. C., and National Research Center, Cairo, Egypt

While aplasia cutis congenita (ACC) as an isolated anomaly has been frequently reported, only 3 reports described the association of ACC with terminal transverse defects (TTD) of the limbs (Adams and Oliver, 1945; Kahn and Olmedo, 1950; and Farmer and Maxmen, 1960). Recently we studied a patient who presented with central skull and scalp defects and TTD of the lower limbs, in addition to which he had both cutis marmorata and markedly dilated and tortuous scalp veins. His mother is clinically normal, but his maternal aunt was said to have similar TTD of the lower limbs. One of the aunt's children, who died at 14 months of age, was noted to have cutis marmorata and dilated scalp veins.

Of the features of this syndrome, ACC as an isolated anomaly has been reported in over 300 cases. Most of these were familial with some suggestive of autosomal dominant inheritance with possible reduced penetrance in some families.

The skull and scalp defects show a wide range of severity varying from minimal scalp defects to deficiency of the bony cranium. ACC is also observed in cases of congenital ring constriction, and with epidermolysis bullosa, focal dermal hypoplasia, D trisomy syndrome and deletion of chromosome 4. These findings illustrate the etiologic heterogeneity of skull and scalp defects of which ACC associated with TTD represents a distinct genetic entity. TTD as an isolated anomaly is a very rare limb malformation which is not simply inherited except in the rare achiropody trait.

The finding of ACC in patients with TTD not associated with congenital ring constrictions is a useful clue to its inheritance. The syndrome of ACC with TTD of the limbs is an autosomal dominant trait with reduced penetrance and variable expressivity. The cutis marmorata, noted in our patient and in previous reports, could be part of the syndrome and could be interpreted as a pleiotropic effect of the mutant gene.

Author Index

Alfi, Omar S., 157
Allderdice, Penelope W., 151
Alvarez, Victor R., 131
Antley, Ray M., 229
Arbisser, Amir I., 219
Arcinue, Edgardo, 209
Atasu, Metin, 279

Bartlett, Catherine, 53
Baum, Jules L., 181
Bixler, David, 229
Blair, John D., 139
Borgaonkar, Digamber S., 87
Bornstein, Paul, 293
Briggs, J., 113
Broome, Diane L., 65
Bull, Marilyn J., 181
Byers, Peter H., 293

Catalano, J. Denis, 39
Chaganti, R. S. K., 169
Chen, Harold, 209
Coldwell, James G., 279
Cox, Hollace L., Jr., 219

Darja, Maria, 187
Davis, Jessica G., 235
Davis, John R., 137
Day, Donald W., 45
Derencsenyi, Anna, 157
Donnell, George N., 157
Ducasse, C., 113
Dumars, Kenneth W., 97
Duncan, Peter A., 275

Ebbin, Allan J., 65

Farnsworth, Peter B., 275
Feild, Eugene, 279
Feinauer, L. Richard, 65
Felismino, Emerita, T., 151
Fialko, Gayle, 97
Fowler, Cynthia G., 315

Geist, Susana, 187
German, James, III, 169
Giles, Harlan R., 137
Goldberg, Michael J., 255
Gooch, W. Manford, III, 161
Gorlin, Robert J., 239, 243
Graap, Raymond, F., 137

Hall, Bryan D., 299
Hall, Judith G., 33, 293
Hamerton, John L., 113
Hardy, Janet B., 23
Harrah, Lyle M., 53
Heine, M. Wayne, 137
Heneghan, Walter D., 151
Herrmann, Jürgen, 247
Hirschmann, Asna, 45
Hoffman, William H., 209
Hood, Carol, 161
Howell, R. Rodney, 219

Idemoto, Judi Y., 267
Israel, Jeanette N., 45
Ives, Elizabeth J., 187

Jackson, Jennifer, 315, 321
James, L. Stanley, xx
Jenkins, Edmund C., 169
Jorgenson, Ronald J., 287
Jung, August L., 65

Lacassie, Yves E., 87
Lafer, Charlotte, 235
Larson, Eunice, 97
Larson, Steven A., 125
Lefkowitz, Martin, 275
Lightner, Elmer S., 137

Madsen, Michael, 65
Mankinen, Carl B., 131
Martens, Paula, 161
Maurer, William F., 267
McAlpine, P. J., 113
Mease, Alan D., 193

Merenstein, Gerald B., 193
Merz, Timothy, 19
Monteleone, James A., 39
Murphy, Edmond A., 7

Narayanan, A. Sampath, 293

Oakley, Godfrey P., Jr., 1
O'Donnell, James J., 299
Ong, Poen S., 219
Opitz, John M., 247

Pallister, Philip D., 247
Pashayan, Hermine M., 255
Pena, Sergio D. J., 113, 201
Perrin, Jane C. S., 209, 267
Pettett, Gary, 193

Rappaport, Elizabeth B., 243
Ray, M., 113
Reed, Joseph O., 209
Reed, William B., 327
Riccardi, Vincent M., 119, 125

Salinas, Carlos F., 287
Say, Burhan, 279

Schlesinger, Les, 327
Scott, Charles I., Jr., 219
Sears, Joseph W., 131
Shapiro, Lawrence R., 275
Shapiro, Steven D., 287
Shokeir, Mohamed H. K., 201
Sirkin, Shari, 299
Smith, David W., 53
Smith, George F., 45
Sotos, Juan F., 267
Stein, Ronald B., 105
Steinberg, Arthur G., 267
Stoll, Claude, 87
Summitt, Robert L., 161

Thurmon, Theodore F., 315, 321
Towner, Joseph W., 105

Ulstrom, Robert, 243

Warnberg, Larry, 279
Wilbur, Lorraine, 169
Wilroy, Robert S., Jr., 161
Wilson, Miriam G., 105
Wiser, Winfred, 161

Yeatman, Gentry W., 119, 125, 193

Subject Index

Abnormalities
  craniofacial, 313
  osseous, 299–313
Acidosis, lactic, 42
Age
  advanced bone, 209–216
  gestational, 26–28
Agenesis
  renal, 53–57
  sacral, 45–50
Amelogenesis imperfecta, 323
Amniocentesis, 161
Amniotic bands, 77
Analysis, trace element, 219, 223–227
Anencephaly, 78
Aneurysm, dissecting, in pregnancy, 229–230, 232
Angioid streaks, 322
Ankylosis, 193–208
Anomalies
  anorectal, 175
  facial, 201–208
  genital, 175
  iris, 229–234
  pelvic, 53
  Peters, 181–186
  placental, 205
  polytopic, 252–253
  single developmental field, 252
  vertebral, 53
Anosmia, 267–273
Antagonists, folic acid, 83
Anus, imperforate, 53–62
Aortic dissection, 229
Aplasia cutis congenita, 332–333
Aplasia, thumb-radius, 253
Apocrine sweat glands, 247–253
Arthrogryposis multiplex congenita, 193–200
Atrophy, muscular, of the lower limbs, 45
Auditory status, 29

Autosomal
  deletion, 169–173
  dominant disorder, 299
  recessive mode, 278
  recessive inheritance, 201, 318

Birth Defects Monitoring Program, 84
Birth defect surveillance, 1–6
Blood phosphorus, elevated, 323
Borderline intelligence, 279–285
Brachydactyly, 279–285

Camptodactyly, 201–208, 279–285
Carpal bones, 253
"Catalog of Chromosomal Variants and Anomalies," 88
Cataracts
  cortical, 40–42
  subcapsular, 311
Chromosomal variants and syndromes, 89
Chromosomes
  analyses, indications for, 130
  extra, small acrocentric (ESA), 119–123
  metacentric, 97–103
  no. 9, 169–173
  no. 10, 169–173
  no. 11, deletion of, 129
  no. 13, partially monosomic, 161–164
  no. 14, partial trisomy, 119–123
  ring Y, 105–111
Cleft, of the soft palate, 235
Collagen, 295–297
Conradi syndrome, 36
Corneal opacities, 191–186
Craniocarpotarsal dysplasia, 331–332
Craniofacial
  abnormalities, 313
  defects, 65–78

Crosslinking, of connective tissue
    components, 293
Cryptorchidism, 205
Cutis laxa, 293–297

Deafness, congenital, 315–320
Dentition, primary, 81
Dermatoglyphics, 175
Developmental delay, 219–227
Developmental field complex
    (DFC), 252
Diagnostic index, 329
Duhamel anomalad, 53–62
Dwarfism, 243–245
Dysgenesis, gonadal, 137–138
Dysplasia
    bilateral renal, 139–148
    craniocarpotarsal, 331–332
    intraheptic bile duct, 147
    microcystic, 278
    oculodentodigital, 329–330
    pancreatic, 147
    renal, 53

Ear lobes, abnormal, 255–265
Ears, low-set and malformed, 139
Ehlers-Danlos syndrome, 297
Elastin, 295
Embryopathy
    oral anticoagulation, 33–37
    vitamin K, anatagonist, 33–37
    warfarin, 33–36
Encephalocele, 65–78
Enzymatic repair, 19–22
Epiphyses, stippled, 33–36
Epistasis, 8

Facies, unusual, 235–238
Fail-safe mechanisms, 13
Fetal alcohol syndrome, 81–82
Fibroblast cultures, 176

Genitalia
    ambiguous, 169–173
    external, defects of, 54
Gestational age, 26–28

Hair, 219–227
Hammertoes, bilateral, 327
Heterochromatin, centromeric, 169
Heterogeneity, 324
Hexadactyly, postaxial, 253
Hiatus hernia, 275–278
Hydramnios, 205
Hymen, imperforate, 247
Hyperaminoacidemia, 42
Hyperglycemia, 42
Hyperkeratosis, marginal gingival,
    239
Hypertension, pulmonary, 209–216
Hypogonadism, 267–273
Hypoplasia
    pulmonary, 147, 186, 193–200
        201–208
    thumb-radius, 253
    of ulnar rays, 247

Ichthyosis, 267–273
Inbreeding, 321
Incomplete penetrance, 15
Infection
    chronic, 26
    maternal-fetal, 23–30
Inheritance
    autosomal recessive, 201–208, 318
    recessive, 219
Intelligence, borderline, 279–285
Intraheptic bile duct
    dysplasia, 147

Joint laxity, 287–290

Karyotype, 46,XX, del (1) (q43),
    131–135

Lax tolerances, 15
Leg bones, deficiency, 253
Lens, intumescent, 40
Limbus dermoid, 140
Lymphocyte cultures, 176
Lysyl oxidase, 293–297

Mammary glands, 247
Marshall syndrome, 299–313
Maternal-fetal infection, pathogenesis
    of, 25–28
Meningomyelocele, 45
Mental retardation, 135, 267–273
Microcephaly, 135, 275–278
Microcystic dysplasia, 278
Micrognathia, 139
Miscegenation, 321
Monitoring systems, 3
Morphogenesis, incomplete uterine, 54
Mosaicism, trisomy 8, 175
Mulattoes, 321
Multiple hit mechanisms, 15
Myopathy
    congenital, 200
    familial, 200

Nephrotic syndrome, 275–278
Noonan phenotype, 81–82
Nose, hypoplastic, 33–34
Nostrils, anteverted, 158

Oculodentodigital dysplasia,
    329–330
Oligodactyly, 253
Osseous abnormalities, 299–313

Palmoplantar, focal, 239
Pancreatic dysplasia, 147
Panencephalitis, chronic, 29
Parastasis, 8
Patella
    congenital dislocation, 279–285
    recurrent dislocation, 287–290
Phalanges, long middle, 158
Pleiotropic dominant mutation,
    247–253
Pleiotropy, mosaic, 252
Polydactyly, 279–285
    of hands, 253
    preaxial, 253
    ulnar, 255–265
Primitive streak, 58–61
Pseudohermaphrodite, male, 169–171

Pseudo-Rubinstein syndrome,
    330–331
Pulmonary cystic disease, 187–191

Q-Banding, 177

Radiation biology, 19
Radiation-induced malformations, 20
Reflux, renal, 46
"Repository of Chromosomal Variants
    and Anomalies," 93
Retinoblastoma, 164
Rubella
    congenital, 25–29
    maternal, 81
    in the second trimenster of
        pregnancy, 27

Salicylates, teratogenic effects, 82
Sensitivity, diagnostic, 12
Sideness, defects of, 61
Sirenomelia, 53
Situs inversus viscerum, 61
Snow-capped teeth, 321–325
Spectrometer, x-ray energy dispersive
    fluorescent, 219
Spina bifida, 45
Stickler syndrome, 299
Strabismus, 135
Syndactyly, hallux, 255–265
Syndrome
    Conradi, 36
    Ehlers-Danlos, 297
    fetal alcohol, 81–82
    Marshall, 299–313
    nephrotic, 275–278
    pseudo-Rubinstein, 330–331
    Stickler, 299
    Turner, 176
    X-linked, 267–273
    XYY, 177
    9p-, 157–159
    9pter→p22 deletion, 151–155

Talipes equinovara, 45
Teeth, snow-capped, 321–325

Teratogenic effect, 26
Teratogens, 1–6
Threshold phenomena, 10
Thumb, triphalangeal, 253, 279
Tissue bands, aberrant, 65, 77
Toenails, hypoplastic, 235
Total parenteral nutrition (TPN),
    39–41
Translocation
    familial Y-22, 97–103
    reciprocal, 170
    X-16 balanced, 138
Trigonocephaly, 158
Triphalangeal thumbs, 279

Triracial groups, 323
Trisomy
    C, 139–148
    2p22 to 2pter, 89–90
    3p21 to 3p26, 90
    8 mosaicism, 175
    tertiary, 113–117
Turner syndrome, 176
Twinning, monozygotic, 53–62

Vitiligo, 315–319

X-linked syndrome, 267–273